New Windows on the Universe

Advances in multimessenger astronomy

Online at: https://doi.org/10.1088/2514-3433/ac81b3

About the program:

AAS-IOP Astronomy ebooks is the official book program of the American Astronomical Society (AAS) and aims to share in depth the most fascinating areas of astronomy, astrophysics, solar physics, and planetary science. The program includes publications in the following topics:

Books in the program range in level from short introductory texts on fast-moving areas, graduate and upper-level undergraduate textbooks, research monographs, and practical handbooks.

For a complete list of published and forthcoming titles, please visit iopscience.org/books/aas.

About the American Astronomical Society

The American Astronomical Society (aas.org), established 1899, is the major organization of professional astronomers in North America. The membership (~7,000) also includes physicists, mathematicians, geologists, engineers, and others whose research interests lie within the broad spectrum of subjects now comprising the contemporary astronomical sciences. The mission of the Society is to enhance and share humanity's scientific understanding of the universe.

Editorial Advisory Board

New Windows on the Universe

Advances in multimessenger astronomy

Saeqa Dil Vrtilek
Center for Astrophysics | Harvard & Smithsonian, Cambridge, MA, USA

IOP Publishing, Bristol, UK

Multimedia content is available for this book from http://doi.org/10.1088/2514-3433/ac81b3.

ISBN 978-0-7503-3731-1 (ebook)
ISBN 978-0-7503-3729-8 (print)
ISBN 978-0-7503-3732-8 (myPrint)
ISBN 978-0-7503-3730-4 (mobi)

DOI 10.1088/2514-3433/ac81b3

Version: 20221101

AAS–IOP Astronomy
ISSN 2514-3433 (online)
ISSN 2515-141X (print)

British Library Cataloguing-in-Publication Data: A catalogue record for this book is available from the British Library.

Published by IOP Publishing, wholly owned by The Institute of Physics, London

IOP Publishing, No. 2 The Distillery, Glassfields, Avon Street, Bristol, BS2 0GR, UK

US Office: IOP Publishing, Inc., 190 North Independence Mall West, Suite 601, Philadelphia, PA 19106, USA

Cover image: A New View of the Tarantula Nebula. Image credit: X-ray: NASA/CXC/PSU/L. Townsley et al.; Optical: NASA/STScI; Infrared: NASA/JPL/PSU/L.Townsley et al.

To the memory of Philip Morrison (1915 – 2005),
who first sparked my interest in astronomy.

Contents

Acknowledgement

I would like to thank William Frucht, who first gave me the idea of writing this book, for starting me on the project, and for editorial comments which greatly enhanced early chapters. I am grateful to IoP editor Leigh Jenkins for gentle nudging that saw the book through completion and to Bhaskaran Balaji for merciless editing of, and astute questions on, early versions of the first five chapters. I am indebted to numerous students and colleagues throughout the years for questions and discussions leading to much of the current text.

Author biography

Saeqa Dil Vrtilek

Dr. Saeqa Dil Vrtilek is currently a Research Associate of the Center for Astrophysics | Harvard & Smithsonian where she served as an Astrophysicist for over 30 years. She has an undergraduate degree in Physics from MIT and a PhD in Astronomy from Columbia University. She is a fellow of the American Association for the Advancement of Science. She has been awarded a Marie Curie Fellowship by the American Association of University Women and a Science Fellowship by the Radcliffe Bunting Institute. Her areas of interest include the physics of accretion disks and jets, multi-wavelength studies of X-ray binaries and planetary nebulae, science education, and public outreach.

Chapter 1

Messengers of Astrophysics

Very Large Array Radio Telescope with clouds. Courtesy of NRAO/VLA.

1.1 Introduction

In the last few decades our view of the universe has changed dramatically. This change is in large part due to our ability to observe the sky with multiple channels. While we can reach some parts of the solar system with probes, outside of the solar system celestial objects are at distances so vast that the only thing we can do to understand them is to study the signals that they emit when nature has found a way to release energy. Energy can be released as electromagnetic radiation, cosmic rays, neutrinos, or gravitational waves. Not surprisingly, these are related to the four fundamental forces of nature: electromagnetic force, strong force, weak force, and gravitational force (Meszaros et al. 2019).

The processes that release the energy are distinct for each messenger and reach the astronomer by distinct means. Humans "see" an object when the electromagnetic

doi:10.1088/2514-3433/ac81b3ch1

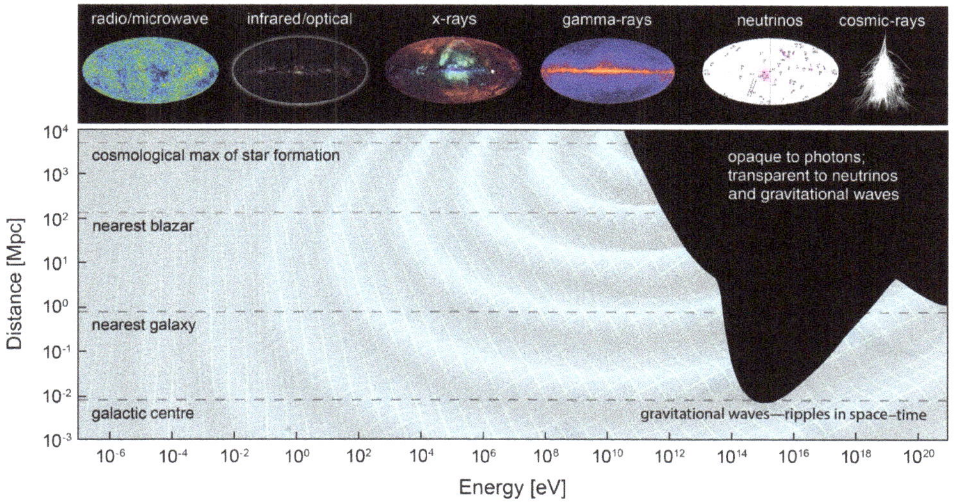

Figure 1.1. Distance horizon at which the universe becomes optically thick to electromagnetic radiation. While lower-energy photons can travel to us from the farthest corners of the universe, the highest-energy photons and cosmic rays are attenuated after short distances, obscuring our view of the most energetic cosmic events. In contrast, the universe is transparent to gravitational waves and neutrinos, making them suitable probes of the high-energy sky. (Reproduced courtesy of Bartos and Kowalski, Institute of Physics Publishing.)

radiation it emits or reflects interacts with cells in our eyes. Astronomers "see" an object when the energy it releases interacts with our detectors. Electromagnetic radiation can be stopped by intervening matter, and cosmic rays interact with matter, but gravitational waves and neutrinos find matter and magnetic fields transparent, hence they allow us to penetrate further into the universe (Figure 1.1; Bartos & Kowalski 2017). Our view of the physical reality of the universe no longer depends on a single mode of seeing but rather on the integration of many different views, all of which, taken together, give us a far deeper understanding of the universe than any one view by itself.

A single book cannot begin to do justice to all that multi-wavelength and multimessenger studies have done to further our understanding of the universe. Instead, I will concentrate on a few examples that demonstrate the tremendous gain in our understanding when we are able to integrate data from different wavelengths and channels into a coherent intellectual structure. These examples represent some of the most exciting areas of astrophysical research, including such diverse topics as the formation and evolution of stars and black holes, and the determination of the Hubble constant, which in effect defines the size, the age, and the content of our universe.

1.2 Electromagnetic Radiation

1.2.1 Covering the Full Spectrum

Electromagnetic radiation is the transmission of energy through space without the need for a physical medium or physical transfer of matter. Energy is transmitted as a massless packet of energy which corresponds to oscillating electric and magnetic fields (also known as a light wave or a photon (Figure 1.2)). Electromagnetic radiation has been the principal way that we have gathered information about our universe. Historically people have watched stars because the nuclear energy release that powers normal stars emerges as visible light off their surfaces. Visible light is the most familiar kind of electromagnetic radiation, but energy can come out in any part of the electromagnetic spectrum, not just that of visible light.

Electromagnetic radiation gives us information about the temperature, density, and composition of the material emitting it. It spans over 20 orders of magnitude in energy from radio waves to gamma (γ) rays (Figure 1.3). All of these are fundamentally similar in that they carry energy and, in a vacuum, move at the same speed, the speed of light in a vacuum (a universal constant that is about 300,000 km or 186,000 miles per second). The difference between them is in wavelength. The shorter the wavelength of the radiation, the higher the energy carried by each "discrete packet" or photon of radiation. We also refer to the "frequency" of radiation, the number of times that the peak of a wave of certain size passes by in a second; hence, the shorter the wavelength the greater the frequency.

Following Karl Jansky's first detections of radio emission from "extraterrestrial sources" (Janksy 1933, 1935), and the first detection of X-rays from the Sun

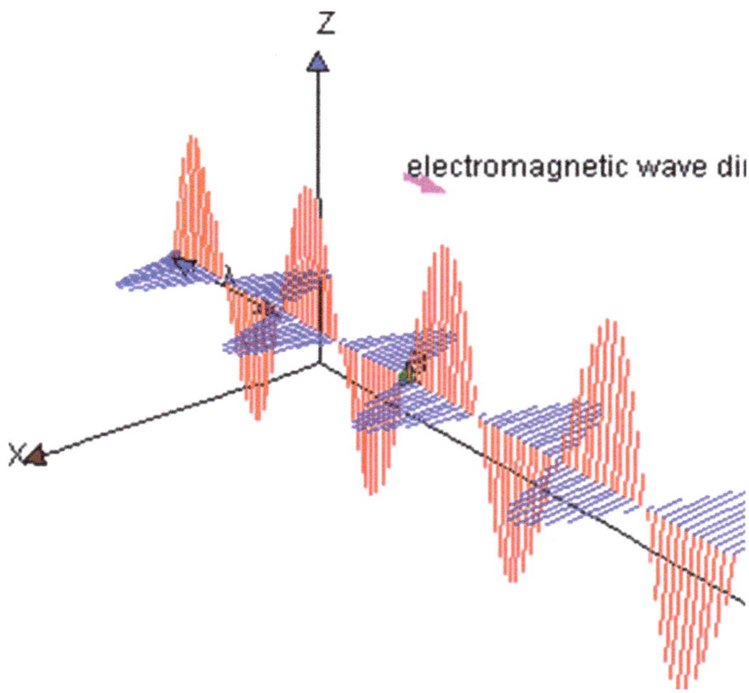

Figure 1.2. An animation showing the direction of propagation of an electromagnetic wave. Red depicts magnetic fields and blue electric fields. These oscillate orthogonal to each other. The direction of propagation of the photon is orthogonal to both. (Reproduced courtesy of Hwang and Esquembre.) Animation available at http://doi.org/10.1088/2514-3433/ac81b3.

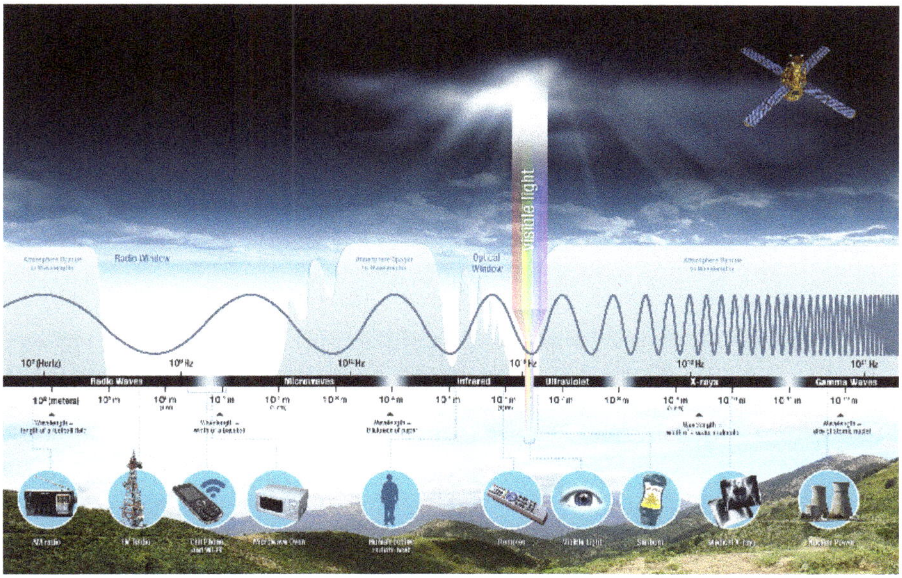

Figure 1.3. This artist's interpretation illustrates that the visible light that we all can see is a very small part of the electromagnetic spectrum. The extremes of the spectrum range from radio waves with wavelengths billions of times longer than those of visible light to gamma-rays with wavelengths millions of times smaller than those of visible light. Familiar objects that give roughly the size of the waves (the distance from crest to crest) are depicted in the figure. While the optical region is rather well defined by what our eyes can see there is some ambiguity in the definition of a regime as ultraviolet or X-rays. The divisions occur because a wide range of technologies is required to look at different bands. The Earth's atmosphere is very good at blocking light from outer space coming down to the surface. As a result, from Earth we can see outer space only at very specific wavelength ranges, called "windows." Radio and visible light fall within such windows, as do a few infrared wavelength ranges. The gray shaded region schematically depicts the height above the Earth's surface that an instrument must be in order to detect the radiation at that wavelength. (Reproduced courtesy of NASA.)

(Friedman et al. 1951), the development of radio telescopes and access to space freed astronomers from the narrow visible, or optical, band of the electromagnetic spectrum. Radio observations have given us many surprises: the detection of pulsars, intergalactic gas clouds, quasars, and the first clear view of our own Galaxy, the Milky Way, among them. Equally spectacular revelations came when the increasing profusion of satellite-mounted instruments permitted observations at wavelengths that cannot penetrate the obscuring effects of Earth's atmosphere (Figure 1.3). The revolution in modern astronomy has been driven to a great extent by the ability to compare observations of celestial sources in different wavelength bands.

Because observing in each of the bands requires complex and unique instruments, the different bands have traditionally been studied by different specialists. A generation ago the great majority of observational astronomers accurately described themselves as either optical, radio, or X-ray astronomers. Hence one needed to piece together information gathered by different astronomers, at different times, and at different wavelengths, into coherent pictures. For long-lived and steady phenomena this approach is feasible, but not all phenomena are long-lived or steady, and throughout the history of astronomy, observations of the time variation of celestial objects have yielded exciting new discoveries and fundamental insights into the nature of cosmic systems. The work of theorists in integrating such observations sometimes led to spectacular solutions of persistent problems. In other cases it led to intriguing new problems.

Beyond the difficulty in coordinating observations in the different bands is the difficulty presented by the large and complex data streams typically obtained at each wavelength. The analysis of the large data streams generated by a modern observatory at even one wavelength can be daunting: a single image can contain gigabytes of information. Astronomers often have to compare and analyze several large data streams at once. Fortunately, another revolution has come to our aid. The tremendous advances in computer technology and the development of standardized data formats and the software to handle them have played a critical role in advancing modern astrophysics. We no longer need two years to construct a pleasing and accurate display of the sky (Figure 1.4). Having many standardized data reduction systems means that every observer does not also have to be a programmer in order to reduce their data.

As we gain ever more information about ever-widening bands of the electromagnetic spectrum, we have to devote increasing effort to integrating information from these disparate sources into coherent explanations of what we see. This is not

Figure 1.4. A picture of the entire sky in visible light made at the Lund Observatory in the 1950s. This is actually a painting made from a composite of photographs: in the 1950s we did not have the computer techniques that enabled us to make one pleasing picture from a collage. The engineers Martin and Tatjana Keskula, well known for their drafting skills, were commissioned by Knut Lundmark to make a painting that would provide an aesthetically pleasing display of the Milky Way. It took the Keskulas two years to complete this painting: there are about 7000 stars in it—the full complement of naked-eye stars over the entire sky—and the features are accurate to the extent that for the individual stars the size reflects relative stellar brightness. The plane of the Milky Way is across the middle of the picture and what we see is starlight with thick lanes of dust across the plane. (Reproduced courtesy of Lund Observatory, Sweden.)

always easy. Each time a band becomes newly available, or undergoes a major technical advance, we uncover new, unexpected, and sometimes seemingly contradictory phenomena. Studies in the infrared, ultraviolet, X-rays, and γ-rays have radically revised the ways we've thought about objects and questions based on visible-light and radio astronomy.

1.2.2 The Resolution: Seeing the Details

The revolution in astronomy is not simply due to the fact that we have opened new eyes—we are also looking farther and sharper with each eye. The angular resolution, or ability of a telescope to distinguish fine detail in the sky increases with the size or diameter of the observing instrument and decreases with the size of the wavelength in which the observation is being made (Figures 1.5–1.8). Our sensitivity, or how far we can see is increased both by increasing the size of the telescope (to collect more light) and also by looking longer (our eyes only see instantaneously but a camera can be exposed for a long time to add up photons from weak and distant sources). Figure 1.9 shows an image made by looking for the equivalent of 11 straight days with the 1.2 m Hubble telescope (although much larger optical telescopes exist on the ground, such deep exposures are not possible because of atmospheric interference). More than 5500 galaxies are visible in this exposure.

The lens of a human eye is roughly 10 mm in length. In Galileo's time the optical telescope (with a lens of diameter 10 cm) for the first time resolved the moons of Jupiter, the rings of Saturn, and individual stars in the Milky Way. Today, with optical lenses greater than 10 meters in size, we study light from the farthest reaches

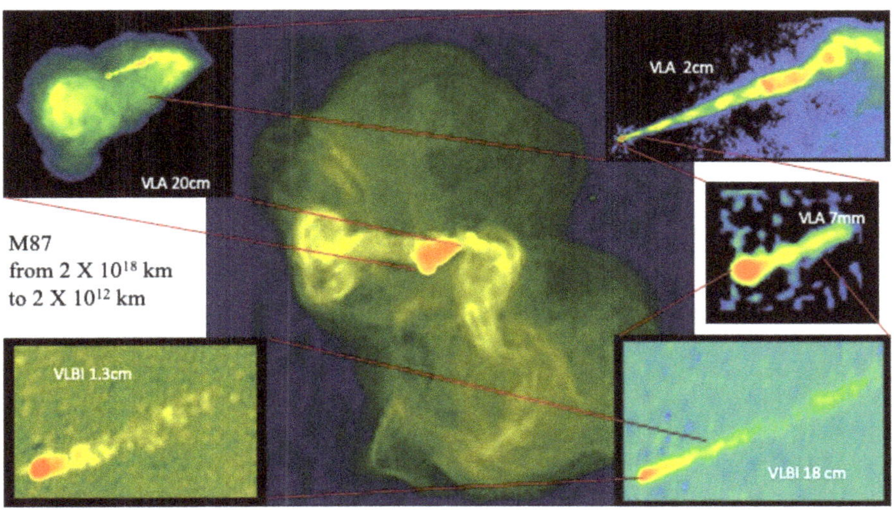

Figure 1.5. Radio images of the quasar M87 at increasing resolution with the lowest resolution using the Very Large Array (VLA) at 90 cm wavelengths (the orange structure indicated by the red lines at the center is over 200,000 light-years (1.9×10^{18} km) across) and the highest resolution using the Very Large Baseline Array (VLBI) at 1.3 cm (1/3rd of a light-year (1.2×10^{13} km) across). (Reproduced courtesy of NRAO/AUI, F. Owen, J. Biretta & J. Eilek.)

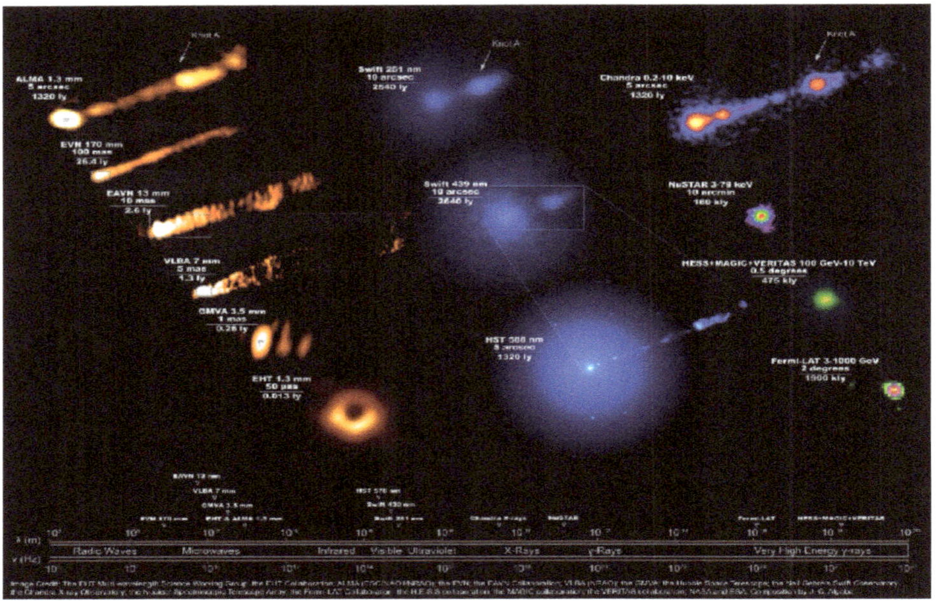

Figure 1.6. The ultimate resolution at radio wavelengths was achieved by the Global Very Large Baseline Array and the Event Horizon Telescope, which captured an image of the black hole at the center of M87 down to 0.013 ly (1.2×10^{14} km). In X-ray and optical wavelengths the highest spatial resolution was reached with the Chandra X-ray and the Hubble Space Telescope reaching around 1300 lys (1.2×10^{16} km). (Reproduced courtesy NRAO/VLA.)

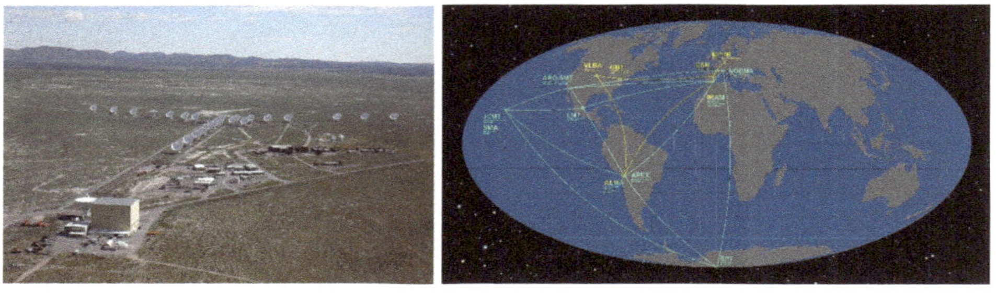

Figure 1.7. Left: (a) The Very Large Array (VLA) consists of twenty-seven 25 m radio telescopes that can be moved along railroad tracks for the closest configuration, which corresponds to a diameter of about 1 km to the most expanded configuration (and highest spatial resolution) spanning a diameter of about 36 km. Located on the high Plains of San Augustin in New Mexico this photograph shows the smallest configuration. (Reproduced courtesy of NRAO/AUI.) **Right (b):** The distribution of telescopes for Event Horizon Telescope (EHT) and the Global Very Large Array mm-VLBI Array (GMVA). The combined use of data from several VLBI clusters around the world combine to form the ultimate interferometer. (Reproduced courtesy of ESO/O. Furtak.)

of the universe. The Hubble ultra deep field shows light that was emitted from galaxies when the universe was less than one billion years old (Figure 1.9). In X-rays we have gone from the handful of bright objects seen in a compilation from the HEAO-1 survey of 1978 (Figures 1.10 and 1.12(a); Gursky et al. 1978), to the

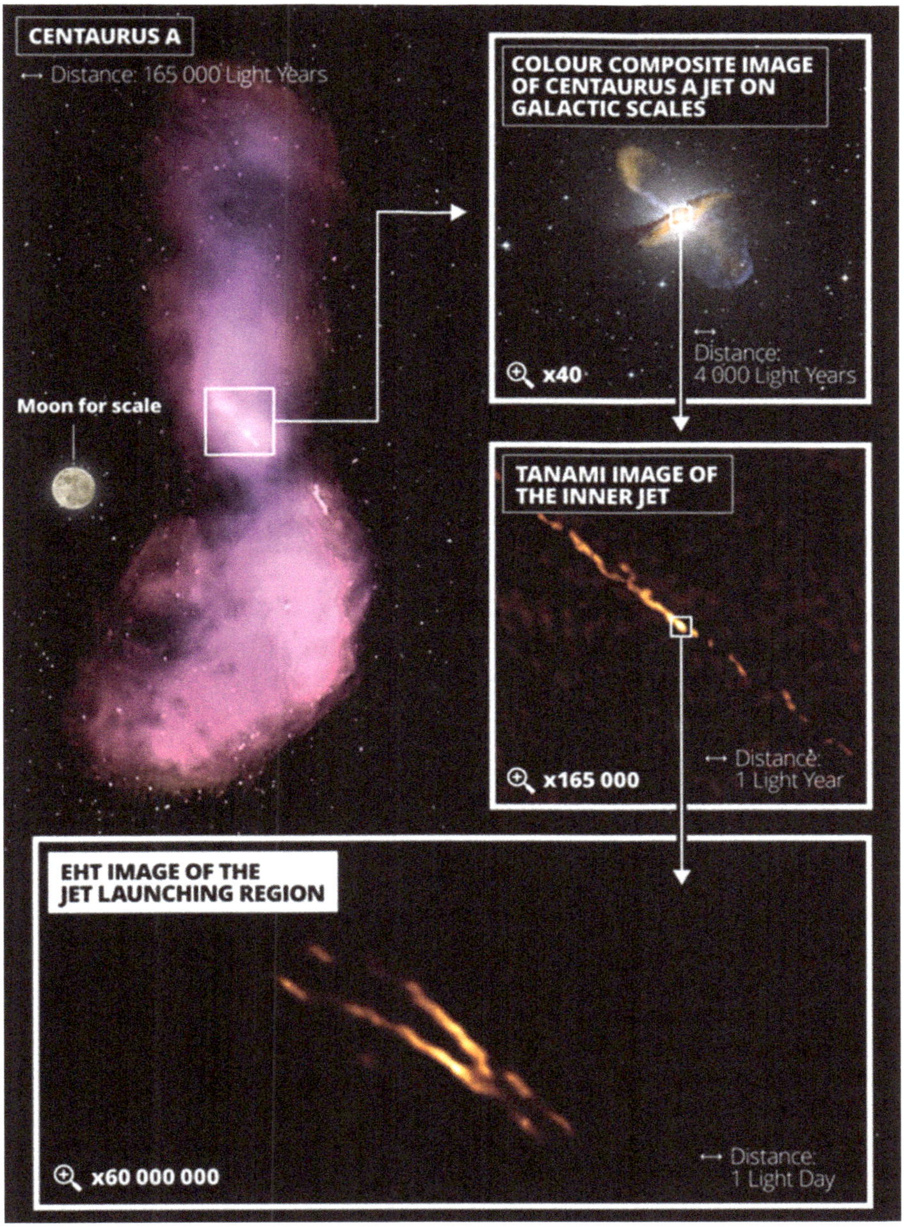

Figure 1.8. The lobes and jets of the active galaxy Centaurus A with the ATCA radio array. Zooming in with X-rays (Chandra blue); submillimeter waves (LABOCA on APEX (orange)); visible in true color with the WFI on the MPG/ESO 2.2 m telescope. With TANAMI Milliarcsecond Interferometry. And the innermost part of the jet obtained with the EHT. (Reproduced courtesy of EHT (Radboud Univ. Nijmegen; CSIRO/ATNF/I. Feain et al., R. Morganti et al., N. Junkes et al.; ESO/WFI; MPIfR/ESO/APEX/A. Weiss et al.; NASA/CXC/CfA/R. Kraft et al.; TANAMI/C. Mueller et al.; EHT/M. Janssen et al.).)

Figure 1.9. The Hubble eXtreme Ultra Deep Field combines 10 years of observations with the Advanced Camera for Surveys and the Wide Field Camera 3, combining 2000 images with a total integration of 2 million seconds, from a small region at the center of the Hubble Ultra Deep Field. The image contains roughly 5500 galaxies within its narrow field-of-view, going back to 13.2 billion years in time. The faintest galaxies are one ten-billionth the brightness that the human eye can see. (Reproduced courtesy of NASA; ESA; G. Illingworth, D. Magee, and P. Oesch, University of California, Santa Cruz; R. Bouwens, Leiden University; and the HUDF09 Team.)

realization that every known class of astronomical objects emits in X-rays as established by the ROSAT satellite in the 1990s (Figure 1.11; Voges 1993).

Examples of improving the focus of our instruments so that we can distinguish really fine features are shown in Figures 1.13 and 1.14. The black and white renditions of the central regions of Andromeda shown in Figure 1.14 demonstrates the incredibly sharp focus of the Chandra X-ray telescope (Garcia et al. 2000; Figure 1.12(b)).

Figure 1.15 shows roughly the time frame of our expansion in wavelength coverage, increase in spatial resolution, and ability to see distant objects. Fifty years ago we could measure only radio and optical radiation; now we have access to essentially all of the electromagnetic spectrum. We currently have operating satellites in nearly all wavelength bands. The dotted lines in Figure 1.15 start at the time of the first detections in the band and the solid lines start roughly at the time that the first detailed observations were made in that band. In almost all bands our resolution is improving rapidly. For a few bright sources radio waves provide the highest resolution by using interferometry techniques (see Chapter 7), but these techniques are now rapidly being extended to shorter wavelengths.

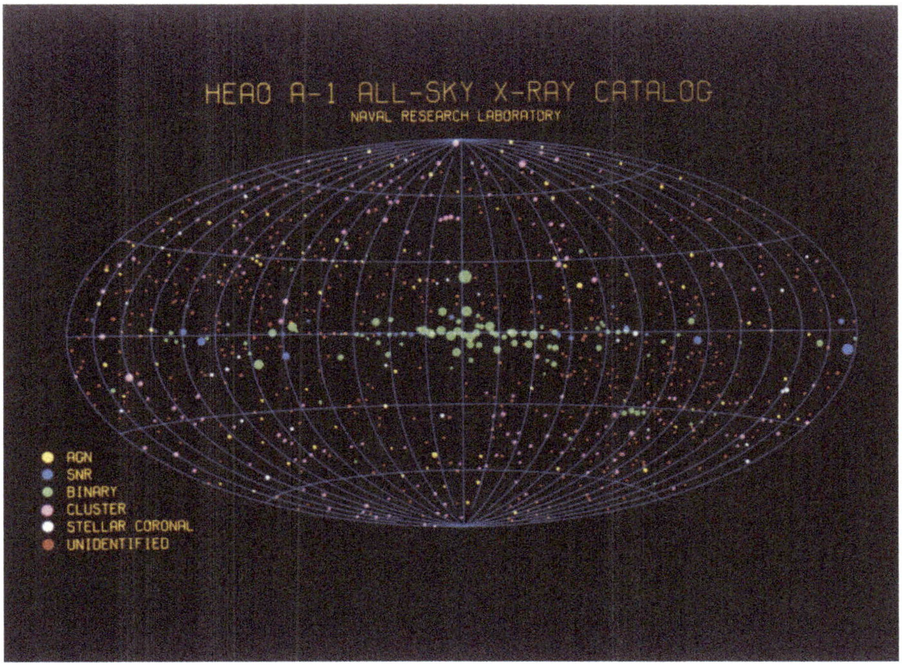

Figure 1.10. An all-sky map made from six months of scanning data from the A2 instrument on board the High Energy Astronomical Observatory (HEAO-1) satellite. (Reproduced courtesy of NASA/GSFC, Keith Jahoda.)

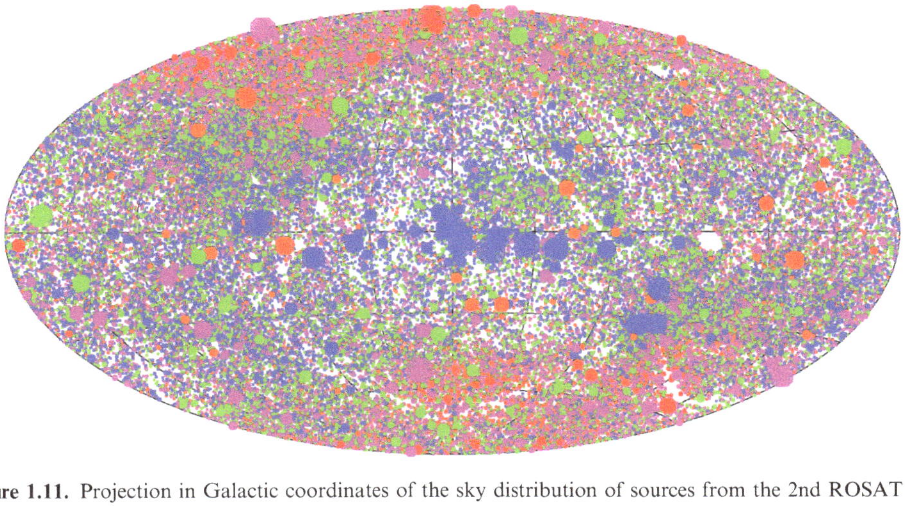

Figure 1.11. Projection in Galactic coordinates of the sky distribution of sources from the 2nd ROSAT All-Sky Survey (2RXS). The size of the symbols scales with count rate, covering values between 0.001 and 68 counts s-1, while the colors (red, magenta, green, blue) represent different hardness ratio ranges: with red indicating softest sources (more low energy photons), while blue sources mark the hardest sources (more high energy photons). The 88,586 objects whose errors on the hardness ratio are smaller than 0.5 are plotted. The faintest sources are detected in the North and South Ecliptic Pole regions, where the exposure time is longest. (Reprinted courtesy of MPE.)

Figure 1.12. Left (a): The HEAO-1 satellite was launched on 1977 August 12 and operated until 1979 January 9. (Courtesy of NASA). **Right (b):** The third of NASA's great observatories, the Chandra X-ray satellite was launched in 1999 and is still in operation as of 2022. (Courtesy of NASA.)

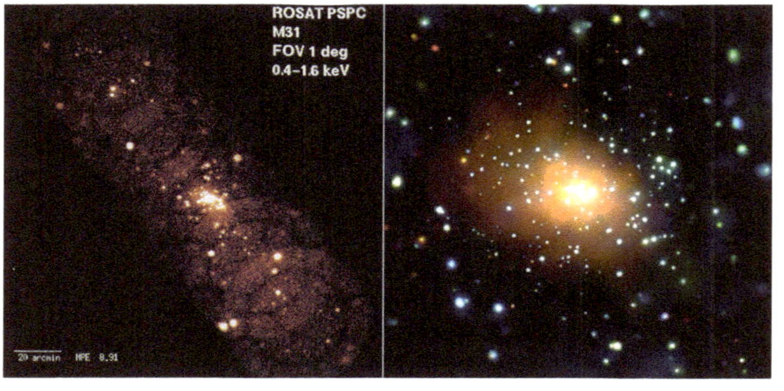

Figure 1.13. Left (a): A mosaic of Andromeda Galaxy observations with the ROSAT PSPC. ROSAT detected as many sources from Andromeda as the first X-ray sky survey by UHURU (1971) detected in the Milky Way. (Courtesy of NASA/Max-Planck-Institut fur extraterrestrische Physik.) **Right (b):** The central regions of the Andromeda Galaxy as observed with Chandra. In addition to resolving hundreds of individual sources, Chandra also discovered a diffuse cloud of hot gas from the central regions. (Courtesy of NASA/UMass/Z. Li & Q. D. Wang.)

1.3 Cosmic Rays

Cosmic rays are actual particles (electrons, protons, and atomic nuclei) that originate outside the solar system and penetrate the Earth's atmosphere. These particles can carry significantly greater amounts of energy than that possible from electromagnetic radiation and enable us to study much higher energy phenomena than can be produced in labs. The numbers of cosmic rays that reach Earth are strongly correlated with their energy (Figure 1.16(a)): the greater the energy, the fewer the number of particles. Each second, about 200 cosmic ray particles with energies of a few million electron Volts (eV) strike every square meter of the Earth.

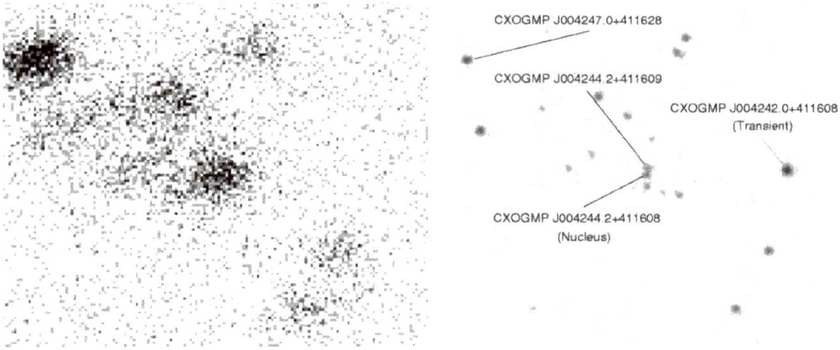

Figure 1.14. Left (a): The center of the Andromeda galaxy from a 84,500 s observation by ROSAT in January 1996. The spatial resolution of ROSAT is 7 arcsec. **Right (b):** The same region in a 8500 s observation by Chandra in 1999; The spatial resolution of Chandra is 0.8 arcsec. The fuzzy patches observed with ROSAT are clearly resolved into individual sources. (Reproduced by courtesy of Garcia et al. 2000.)

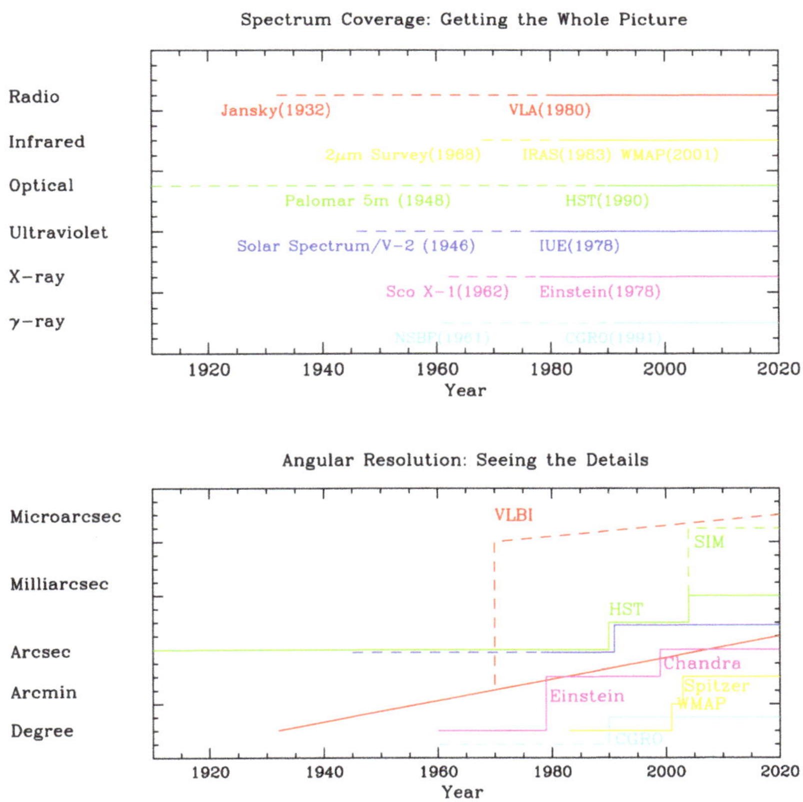

Figure 1.15. Upper panel (a): Schematic representing the year and the instrument that made the first observation in that wavelength. **Lower panel (b):** The smallest angle to be resolved at a given wavelength. One arcsec is 1/360th of a degree as subtended on the sky.

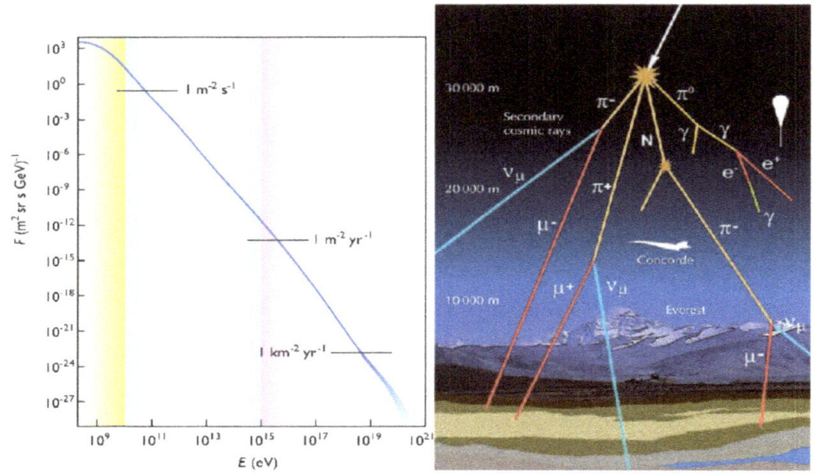

Figure 1.16. Left (a): The frequency of arrival of cosmic rays at Earth's surface as a function of energy. Yellow region is attributed mostly to Solar cosmic rays, blue to Galactic cosmic rays, and purple to extragalactic cosmic rays. (Reproduced courtesy of Sven Lafebre.) **Right (b):** When a cosmic ray particle hits a nucleus in the atmosphere it decays into many subatomic particles: photons, leptons, electrons, positrons, muons, and pions. (Reproduced courtesy of PhysicsOpenLab.)

These are going through you right now. At a billion (10^9) eV there is only one per square meter per second. Above 10^{18} eV there is only one particle per week for each square kilometer. The highest-energy cosmic ray detected as yet had an energy of 3×10^{20} eV, several million times more energetic than the highest-energy gamma-rays recorded (of order 10^{14} eV), and 40 million times more energetic than that achieved as yet by man with the Large Hadron Collider. They are powerful enough to damage electronics in space and, increasingly, microelectronics on Earth and are an important factor to consider when developing life carrying vehicles outside the Earth's atmosphere.

The majority of cosmic ray particles are simple protons (90%); slightly less than 9% are helium nuclei, about 1% are electrons, and there are trace amounts of heavier nuclei. As charged particles, cosmic rays are strongly influenced by the Earth's magnetic field and the Sun's solar wind. At distances of approximately 94 AU from the Sun, the solar wind undergoes a transition, called the termination shock, from supersonic to subsonic speeds. The region between the termination shock and the heliopause acts as a barrier to cosmic rays, decreasing the flux at lower energies by about 90%. The amount of solar wind is not constant due to changes in solar activity. Thus, the level of the cosmic ray flux varies with solar activity. The Earth's magnetic field deflects some of the cosmic rays resulting in an apparent dependence on latitude, longitude, and azimuth angle. When a cosmic ray particle hits a nucleus in the Earth's atmosphere, its energy is dispersed into billions of subatomic particles (photons, leptons, electrons, positrons, muons, and pions) all moving at near the speed of light (Figure 1.16(b)). Subatomic particles moving at near the speed of light generate Cherenkov radiation (named after the person who first detected them;

Cherenkov 1934), which can be distinguished from that caused by very high-energy gamma-rays because in addition to electrons and positrons, they also produce significant numbers of muons, pions, and neutrinos. These secondary particles, originating from a single cosmic ray, are referred to as "air showers." Cherenkov radiation has the property that it can travel within a medium, such as water, faster than light travels in the medium.

Water Cherenkov tanks for detection of neutrinos were first developed in the late 1950s at the University of Leeds by J. G. Wilson (Wilson 1956). An array of tanks covering 12 km^2 were dedicated to detecting particles with energies of 10^{20} eV. When the array closed in 1987 the events that had been detected suggested a flux at this high energy of less than one particle per square kilometer arriving on Earth per century. At these high energies, in order to record a significant number of these events, detectors are needed to cover a large area.

A few of the high energy astrophysicists at Leeds become founding members of the Pierre Auger Observatory, located in the Mendoza province of Argentina. Pierre Auger has 1660 water Cherenkov detectors and 24 fluorescence detectors spread over an area of 3000 km^2 (the size of Rhode Island, USA; Figure 1.17). Pierre Auger was designed to detect cosmic rays in the 10^{17}–10^{20} eV range, termed Ultra High Energy Cosmic Rays (UHECR). Simple geometry using simultaneous observations of air showers from several Cherenkov detectors can be used to determine the direction from which the ray was emitted. The fluorescence detectors were first developed at Volcano Ranch, New Mexico and a higher resolution version was developed at the University of Utah (Figure 1.18(a)). Called Fly's Eye because they use faceted mirrors to produce pictures at high speed, they detect fluorescence of the nitrogen molecules, which appears in ultraviolet light, as the shower particles traverse the sky. This also provides information on the direction of the original cosmic ray particle. Pierre Auger is the first observatory to utilize both types of detectors at the same location, enhancing the precision with which directions can be measured.

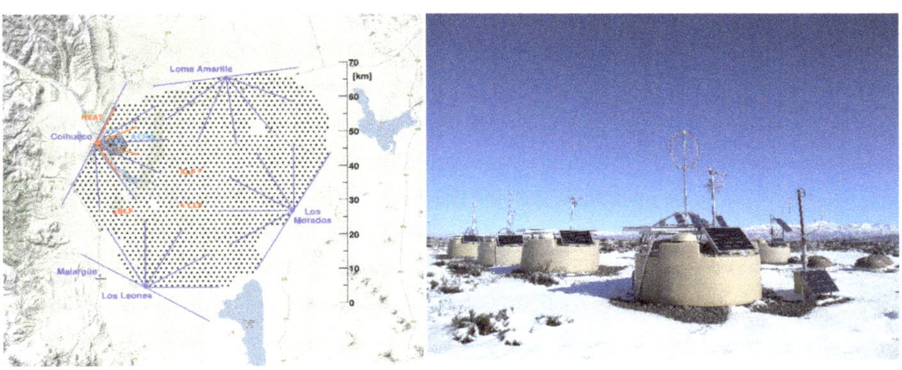

Figure 1.17. Left (a): The Pierre Auger Observatory in Mendoza Province in western Argentina consists of 1660 water Cherenkov detectors (gray dots) and 27 telescopes designed to detect fluorescence (blue rays). **Right (b):** One of the 1660 water Cherenkov detectors at the Pierre Auger Observatory. The tanks are about 4 ft in height with a diameter of 8 ft. (Reproduced courtesy of Pierre Auger Collaboration.)

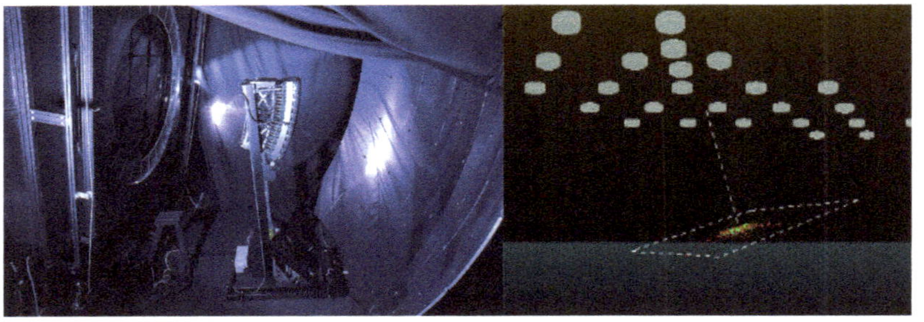

Figure 1.18. Left (a): Two of the High Resolution Fly's Eye cosmic ray telescopes, developed at the University of Utah and serving as models of the fluorescence detectors at Pierre Auger. (Reproduced courtesy of Pierre Auger Observatory.) **Right (b):** In this animation the air shower particles are illustrated using green dots to represent electrons and positrons, and red dots to represent muons. Yellow indicates that particles have been detected coincidentally by one of the water tank detectors. Slight differences in the detection times at the various tank positions allow scientists to determine the cosmic ray arrival direction. (Reproduced courtesy of Pierre Augere Collaboration.) Animation available at http://doi.org/10.1088/2514-3433/ac81b3.

1.4 Neutrinos

Neutrinos are subatomic particles with masses much less than that of the electron; because they can travel mostly unaffected by matter, they provide a natural probe to cosmological distances. Neutrinos also carry no charge hence they are not affected by electromagnetic forces and reliably point back to their source of origin.

Neutrinos are formed during radioactive decay and in nuclear reactors but are also created in several astronomical contexts: in nuclear reactions that take place in the centers of stars (about 65 billion solar neutrinos per second pass through every square centimeter on Earth that is perpendicular to the Sun), when cosmic rays hit atoms, and in supernovae (a single supernova event releases an octodecillion (10^{57}) neutrinos). The neutrinos lack of reaction to electromagnetic forces also means that they cannot be detected directly. The weak interaction of neutrinos with matter (neutrinos generated in the center of the Sun pass through the Sun without interaction and the surface area of the Earth opposite the Sun receives almost the same number as the side facing the Sun) make them very difficult to detect.

Wolfgang Pauli first suggested the existence of neutrinos in 1930 in order to conserve energy and momentum observed in radioactive decay. The first experimental detection of a neutrino interacting with matter occurred in 1956 at the Savannah River Plant in South Carolina (Cowan et al. 1956; Reines & Cowan 1956). In the 1960s John Bahcall and collaborators predicted the number of neutrinos the Sun should produce (Bahcall et al. 1963). The majority of solar neutrinos are low-energy: direct products of proton–proton reactions with energies up to 400 keV. Based on Bahcall's prediction Raymond Davis updated an earlier version of an experiment he had designed to detect neutrinos from the Sun. Bahcall's prediction and Davis's experiment were described in separate papers published side by side (Bahcall 1964; Davis 1964). Neutrinos can also cause Cherenkov radiation and in order to separate events caused by neutrinos with those caused by Cosmic Rays, Davis buried the detector (named Homestake for the town in South Dakota it was built in) deep underground where Cosmic Rays cannot penetrate. The Homestake experiment found a neutrino flux from the Sun that was one-third the value expected from theory (Davis et al. 1968; Bahcall and Davis 1976); the discrepancy between prediction and detection remained a mystery for over 20 years.

In 1983, the Japanese completed the Kamioka Nucleon Decay Experiment (Figure 1.19(a)) which was a cylindrical tank about 16 m high and 15.6 m wide, contained 3058 metric tons of pure water, lined with 1000 photomultiplier tubes (PMTs), and buried in a mine 1006 m underground. In 1985 they added an additional shell around KamiokaNDE in order to shield the experiment from gamma-rays from surrounding rock. Kamiokande-II operated from 1985 to 1990 and was the first experiment to detect non-solar neutrinos with the observation of 11 neutrino events ranging from 10 to 40 MeV from Supernova 1987A (see Chapter 2). Kamiokande-II detected a flux of neutrinos one-half that of the theoretical expectation from the Sun. The first experiments to solve the discrepancy between observed and detected solar neutrinos, the Super-Kamiokande (Figure 1.20), was completed in 1996. With 50,000 metric tons of ultra-pure water and 13,000 PMTS, Super-Kamiokande discovered neutrino oscillation, an effect which causes neutrinos to oscillate between three types: electron, muon, or tau (Fukuda et al. 1998). This change of form accounted for the discrepancy. Homestake was able to detect only one of the three neutrino types. The Super-Kamiokande was also the first detector to image the Sun in neutrinos (Figure 1.19(b)).

IceCube is an experiment built at the Amundsen-Scott South Pole Station in Antarctica under the supervision of the University of Wisconsin-Madison and involving 250 physicists and engineers from 11 countries (Figure 1.21). Completed in 2011, IceCube takes advantage of Digital Optical Modules (DOMs) each with a photomultiplier tube (Figures 1.22(a) and (b)). DOMs work on the same principle as

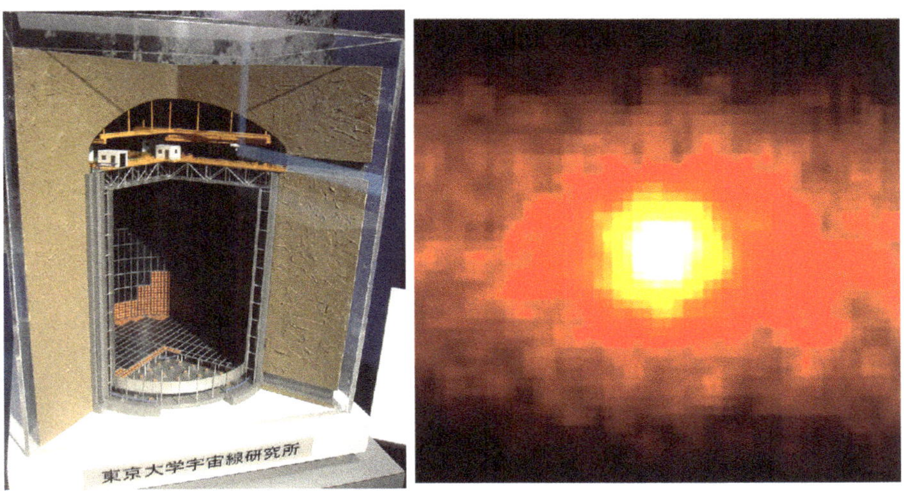

Figure 1.19. Left (a): A model of the Kamioka Neutrino Detector. Located 1000 ft underground it held 3058 metric tons of pure water and was lined with 1000 photomultiplier tubes. (Courtesy of Kamioka Observatory, ICRR). **Right (b):** Our Sun as imaged in neutrinos by Super-Kamiokande. (Reproduced courtesy of R. Svoboda and K. Gordan (LSU).)

Figure 1.20. Super-Kamiokande held 50,000 gallons of pure water and had over 11,000 photomultiplier tubes. This image shows the photomultiplier tubes being cleaned by workers in a raft. (Courtesy of Kamioka Observatory, ICRR, The University of Tokyo.)

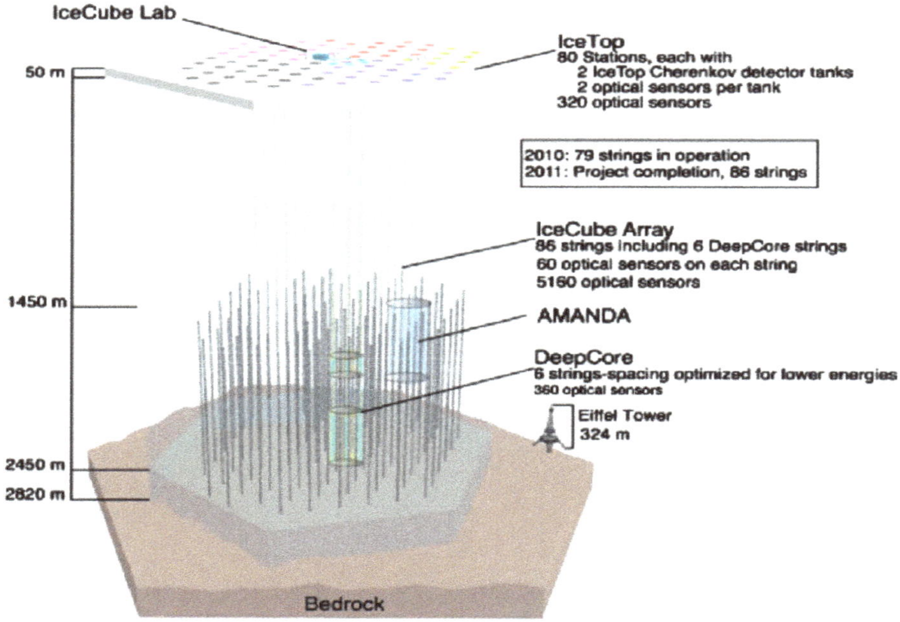

Figure 1.21. A schematic showing the design of IceCube upon completion in 2011. 5160 Digital Optical Monitors strung in 86 boreholes between 1450 and 2450 m below the ice, covering an area of one square kilometer. The Deep Core consists of an additional six strings in the clearest part of the ice with 60 DOMs placed closer together to optimize detection at low energies, and IceTop which has 80 stations each with 2 sensors and 2 Cherenkov detector tanks. The inset labeled AMANDA was the first part built and operated between 2006 and 2009 as proof of concept. A image of the Eiffel Tower is used to show the scale. (Reproduced courtesy of the IceCube Collaboration.)

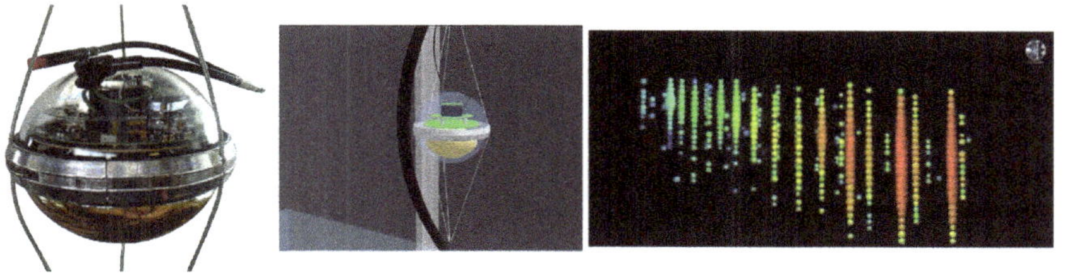

Figure 1.22. Left (a): One of the Digital Optical Monitors (DOM) from IceCube hole 85. Proudly signed by the engineers who built it. **Center (b):** Animation showing how the DOMs (60 for each of 86 strings), are distributed along strings. When a DOM senses energy from Cherenkov radiation it sends it up to the IceCube lab which is on the surface. (Reproduced courtesy of the IceCube Collaboration.) **Right (b):** The event from TXS 0506+056d represented in schematic. Each colored circle represents a DOM that was triggered, with the larger circles denoting brighter events, earlier times in red, later times in green. In this image the particle is moving from right to left. (Reproduced courtesy of the IceCube Collaboration.) Animation available at http://doi.org/10.1088/2514-3433/ac81b3.

water Cherenkov detectors: neutrinos interact with the ice producing secondary charged particles that emit Cherenkov radiation. In 80 boreholes (1450 m to 2450 m deep drilled into the ice with hot water drills) are placed strings each with 60 DOMS spaced 17 m apart. From May 2010 to May 2012 IceCube observed 28 neutrinos with energies above 50 TeV. Reported in 2013, this constituted the first detection of very high-energy neutrinos from outside the solar system. In 2017, IceCube detected a 290 TeV muon neutrino (Figure 1.22(c)). Both space and ground-based observatories around the world were notified within minutes of the detection. By combining information from gamma-ray observatories MAGIC and the Fermi satellite the emission was identified as originating from an active galactic nucleus (TXS 0506+056; IceCube Collaboration, Aartsen et al. 2018). This was the first identification of an extragalactic object as a source of neutrinos. Archived data showed that neutrinos from TXS 0506+056 were also observed in 2014–2015. The fluxes detected by IceCube suggests that the majority of non-thermal energy from the universe is produced in the equivalent of hadronic accelerators in black holes and neutrons stars.

IceCube detects fewer events than more traditional detectors (such as Super-K) but at much higher energies and 3 times higher resolution. It is sensitive to all three types of neutrinos and measured a high background level of high-energy neutrinos (HENs) at TeV to PeV energies. In time it will build up a map of neutrino flux in the northern hemisphere similar to the cosmic microwave background.

1.5 Gravitational Waves

While electromagnetic radiation is produced by moving electric and magnetic fields, gravitational waves are produced by the movement of masses. Gravitational waves were predicted by the general theory of gravitation published by Einstein in 1915. The universal law of gravitation that Newton had developed (based in part on the experiments of Galileo and others) in 1687 had been remarkably effective in explaining the behavior of objects in our solar system and had successfully predicted the presence of Neptune based on perturbations on the orbit of Uranus. The equations developed by James Clark Maxwell in the middle of the 19th century showed that electricity and magnetism were related and worked beautifully to explain all observed electromagnetic phenomenon. However, Newton's laws of mechanics and Maxwell's equations for electricity and magnetism were incompatible. Newtonian/Galilean mechanics required speeds to add: if someone is traveling toward you at 10m s^{-1} and threw a ball toward you at a speed of 10m s^{-1} the ball would travel at 20m s^{-1}; if you turned on a flashlight while traveling in a car the speed of the car should contribute to the speed of the light from the flashlight. Newtonian/Galilean mechanics also required that gravity act instantaneously over large distances, although Newton himself was unhappy with this requirement. Maxwell's equations required that the speed of light be finite and constant; finite means that light takes time to travel from one point to another; constant means no matter how fast you are moving when you turn on a flashlight, the light from the flashlight travels at the same speed. By the end of the 19th century, people had already realized that Newton's laws could not account for all gravitational effects (notably, the perihelion shift of Mercury) whereas Maxwell's equations worked beautifully for everything they were applied to, so Einstein assumed Maxwell was right and opted for a constant speed of light. This, with the further assumption that the laws of physics are invariant in non-accelerating frames, led to his Special Theory of Relativity (SR) in 1905. The consequences of SR are not intuitive, strange things happen: clocks slow down when moving, moving objects shrink in the direction of motion, mass increases. Expanding his theory to include accelerating frames Einstein came up with the General Theory of Relativity (GR) in 1911. In GR space itself warps (Figure 1.23). One of the consequence of having a finite speed of light is that time can be considered analogous to a dimension resulting in a four-dimensional spacetime. Gravity is then a geometric property of spacetime with mass a source of gravitational radiation. And gravitational waves are disturbances in the curvature of spacetime caused by accelerating masses. I tell my students that they can check for themselves the fact that time is relative: when they were four years old one year was 1/4th of their lifetime and a year seemed to last for ever; now they are twenty years old one year is 1/20th of their lifetime and the years go by much faster. At least the concept that it is hard to define time in absolutes is made clear. But how express space? On the surface (pun intended) our sense of space appears easy to define: we can define a standard length and use it to measure all other distances. But Einstein showed that space and time are inexorably related. As beautifully expressed by John Wheeler: "Mass tells spacetime how to curve; spacetime tells matter how to move." (Figure 1.23; Wheeler & Ford 1998).

Einstein's theory is the simplest theory of gravitation that explained phenomenon that could not be understood using classical physics, such as the perihelion shift of Mercury. Newtonian theory of gravitation predicted a shift of 531 arcsec per century due to the effect of other planets on Mercury's orbit around the Sun. In the 19th century the shift was measured at 574 arcsec. Einstein's theory of gravitation predicted an additional effect due to the curvature of spacetime around our Sun: the amount was 43 arcsec per century. Einstein's theory also made many other predictions that defied classical physics most of which have since been proven.

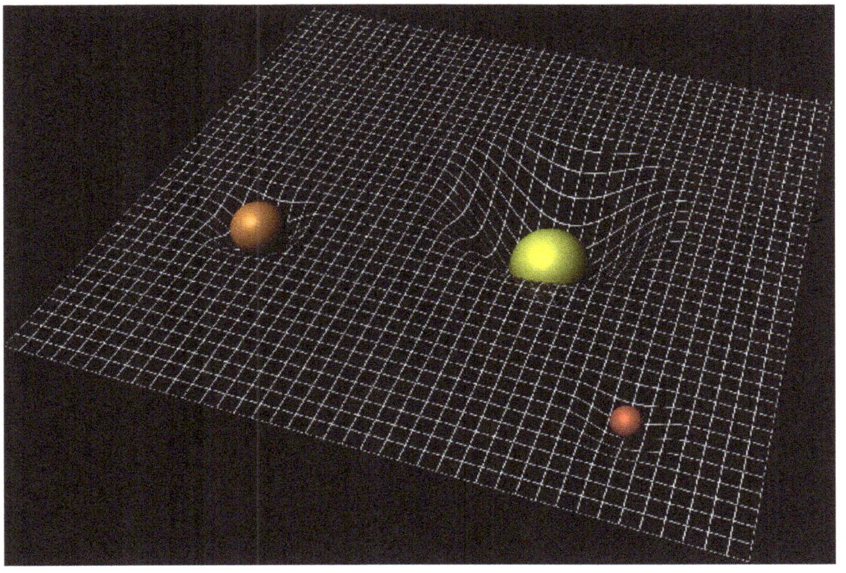

Figure 1.23. A representation of spacetime warping by mass. The greater the mass the greater the "warp." Imagine different mass balls on a flexible sheet. Sheet is in two dimensions and warping is in a third dimension. We are in three-dimensional space so Einstein's gravity and/or acceleration warping will occur in a fourth dimension. (Reproduced courtesy of NASA/LISA.)

The bending of light by gravity was verified as early as 1919 by Sir Arthur Eddington's detection of a star known to be behind the Sun during a solar eclipse (Eddington 1919). The gravitational lensing effect due to the bending of light by gravity has become a major tool in observational astronomy (see Chapter 6), allowing us to see objects at far greater distances than possible with current telescopes, or objects such as planets that are too dark to be detected in their own light. Gravitational redshift was detected in 1959 (Pound & Rebka 1959). Gravitational time dilation and the slowing down of moving clocks are both needed to locate positions and have been used by Global Positioning System (GPS) for years. The first indirect measurement of gravitational waves occurred in 1974 when Hulse and Taylor observed two neutron stars orbiting each other, the two objects were nearing each other at the rate predicted by general relativity due to radiation of energy from the moving masses (Figure 1.24(a); Hulse & Taylor 1975).

However, the first direct detection of gravitational waves did not occur until 2015 a hundred years after Einstein predicted them. This is because the signal to be measured is a change in distance of one part in 10^{21}, one thousandth the diameter of the proton. The discovery of lasers in the 1960s led Rainer Weiss of MIT to suggest laser interferometry as a possible means to achieve such resolution (Weiss 1972). More than forty years later, his initial concept grew into the Advanced Laser Interferometer Gravitational Wave Observatory (Advanced LIGO), which made the first detection with a sensitivity one ten-thousandth the diameter of the proton. The detectors can be considered two of the world's largest vacuum tubes, each consisted

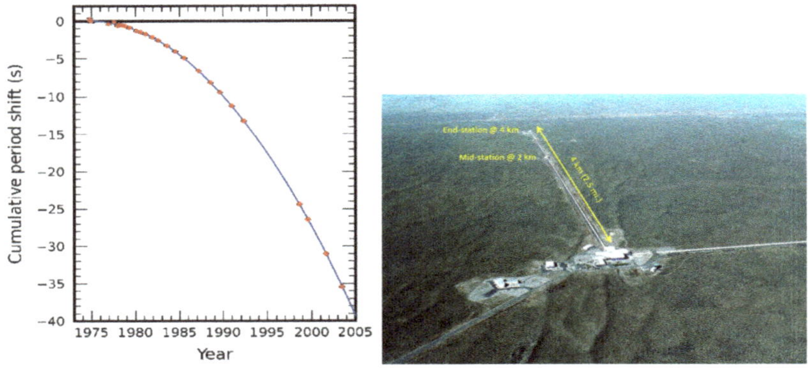

Figure 1.24. Left (a): The red dots represent 32 years of observations of the Hulse–Taylor Binary Pulsar B1913 +16 and the blue line is the theoretical prediction of periastron shift representing energy loss from the system through gravitational radiation. **Right (b):** Aerial view of the LIGO facility in Hanford, Washington. One arm is labeled with the locations and distances to the mid- and end-station. The perspective distorts the distances between them. (Reproduced courtesy of NASA/LIGO.) (Reproduced courtesy of Inductiveload/Wikimedia.)

of a l-shaped tube with two 4.0 km (2.5 mi) legs set at right angles to each other (Figure 1.24(b)). The 3002 km (1865 miles) between the two facilities translates to up to 10 milliseconds in light transit time. If a signal identical to that detected in one is seen in the other within 10 milliseconds it can be safely thought to originate in space rather than some local effect. GW150914 (labeled by its date of detection, 2015 September 14), the first gravitational wave detected was reported by B. P. Abbott for a collaboration of nearly 1000 people (Abbott et al. 2016). It matched the waveform predicted by GR for the merger of two stellar-mass black holes into a single black hole and therefore also identified the existence of binary stellar-mass black holes (Figure 1.25). Figure 1.26 is an animation illustrating how spacetime is distorted as the two black holes near each other. At the moment of collision they are moving at half the speed of light. Accelerators on Earth have difficulty getting an atom-sized mass to move at a fraction of the speed of light! The single black hole that resulted from the collision was 3 solar masses lighter than the combined mass of the two objects. This means that 3 solar masses of energy were released in gravitational waves.

In 2017 August 17, LIGO saw two neutron stars collide (Abbott et al. 2017). The black hole collision lasted a fraction of a second. Whereas the neutron star collision lasted several seconds. The European experiment VIRGO also observed this collision. Having three detectors allowed much better localization of where the event occurred. Observatories at all wavelength in the Southern hemisphere, both space and ground based, turned telescopes toward the direction determined. 1.7 s after LIGO saw the signal the FERMI gamma-ray observatory saw a signal. An optical signal was detected by multiple observatories 11 hours later. An extant image of the same field showed nothing 20 days earlier. The first electromagnetic detection of a gravitation wave detection was established. Even more amazing, the observed spectrum showed lines that could be associated with specific elements, including heavy ones such as gold. Our Sun does not produce heavy elements, and heavy elements released by supernovae explosions did not produce the quantity of gold

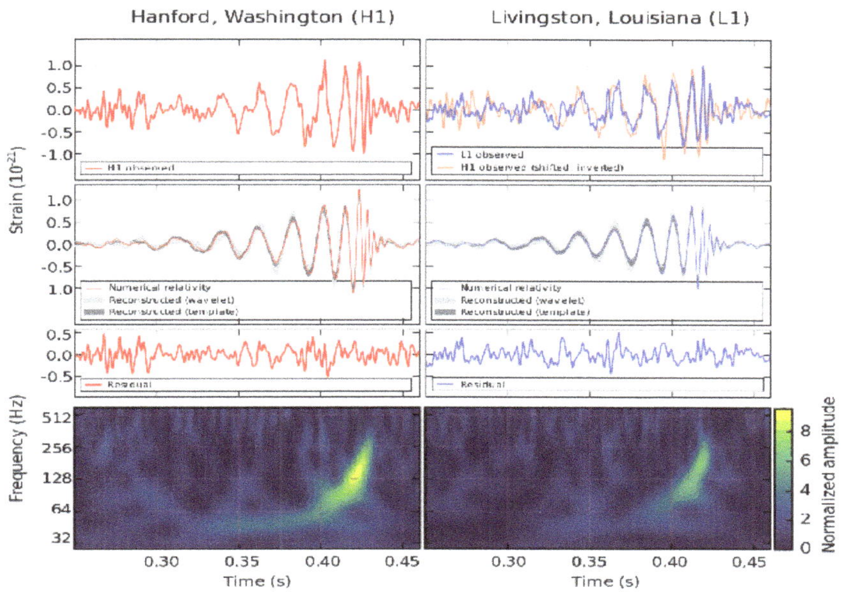

Figure 1.25. The first confirmed detection of a direct signal from gravitational waves. The Livingston, Louisiana signal arrived 7 milliseconds after the signal detected in Hanford. Horizontal axis is time in section vertical axis is strain (in units of 10^{-21}). Amplitude is 10^{-21} at maximum, mirror movement displacement is about 4×10^{-18} m. The total duration of the event was a fraction of a second. (Reproduced courtesy of LIGO Collaboration.)

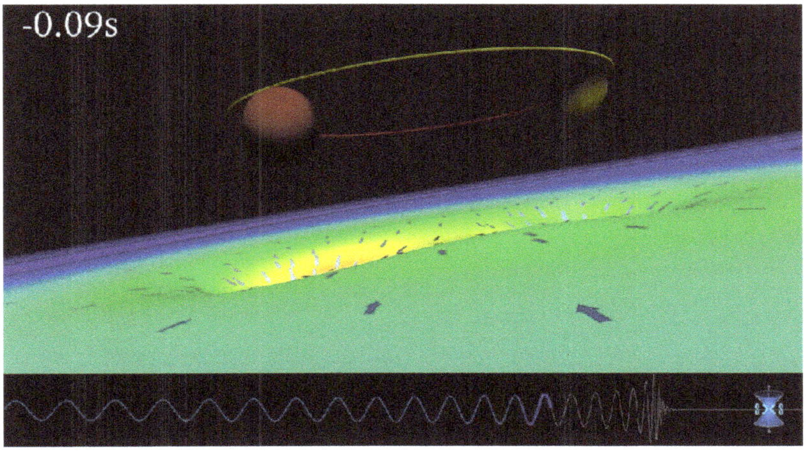

Figure 1.26. A computer simulation depicting the distortion of spacetime by two black holes orbiting each other as observed by LIGO on 2015 September 14. The black holes with inferred masses 29 and 36 times that of our Sun give off gravitational waves as they get closer and closer together. The combined black hole has a mass of 62 solar masses. (Reproduced courtesy of LIGO/Caltech.) Animation available at http://doi.org/10.1088/2514-3433/ac81b3.

(relative to other elements) that we find on Earth. Collisions of neutron stars may produce the amounts needed.

This was a tremendous achievement in just two years with just two detections. Twenty-two gravitational events have been detected by LIGO and VIRGO as of April 2021; but this is not nearly enough to provide statistics on the many types of events possible. Many astrophysical objects from supermassive black holes in galactic nuclei to compact binaries in our own Galaxy, are sources of gravitational waves. Frequencies that will enable us to study objects beyond our Galaxy are still being developed.

The direct detection of gravitational waves has opened one of the last windows left with which to explore our universe. As we know, the more ways we have of looking at something, the more opportunity there is for unexpected connections and ideas. Studying astrophysical phenomena in ways that encompass all available channels promises both greater understanding and the possibilities of greater mysteries in the next few decades.

References

Abbott, B. P., Abbott, R., Abbott, T. D., et al. 2016, ApJ, 818, 22

Abbott, B. P., Abbott, R., Abbott, T. D., et al. 2017, PhRvL, 119, 161101

Aartsen, M.G., Ackermann, M., Adams, J., et al. 2018, Sci, 361, 1378

Bahcall, J. 1964, PhRvL, 12, 303

Bahcall, J. N., & Davis, R., Jr. 1976, Sci, 191, 264

Bahcall, J., Fowler, W. A., Iben, I. Jr+, & Sears, R. L. 1963, ApJ, 137, 344

Bartos, I., & Kowalski, M. 2017, Multimessenger Astronomy (Bristol: Institute of Physics Publishing)

Cherenkov, P. A. 1934, Proceedings of the USSR Academy of Sciences (Russian: Доклады Академии Наук СССР, Doklady Akademii Nauk SSSR (DAN SSSR), 2, 451

Cowan, C. L. Jr., Reines, F., Harrison, F. B., Kruse, H. W., & McGyire, A. D. 1956, Sci, 124, 103

Davis, R. 1964, PhRvL, 12, 303

Davis, R., Harmer, D. S., & Hoffman, K. C. 1968, PhRvL, 20, 1205

Eddington, A.S. 1919, Natur, 104, 372

Friedman, H., Lichtman, S. W., & Byram, E. T. 1951, PhRv, 83, 1025

Fukuda, Y., Hayakawa, T., Ichihara, E., et al. 1998, PhRvL, 81, 1562

Garcia, M. R., Murray, S. S., Primini, F. A., et al. 2000, ApJ, 537, L23

Gursky, H., Bradt, H., Doxsey, R., et al. 1978, ApJ, 223, 973

Hulse, R. A., & Taylor, J. H. 1975, ApJ, 195, 51

Jansky, K. G. 1933, Proc. Institute of Radio Engineers, 21, 1387

Jansky, K. G. 1935, Proc. Institute of Radio Engineers, 23, 1158

Meszaros, P., Derek, B. F., Hanna, C., & Murase, K. 2019, NatRP, 1, 585

Pound, R. V., & Rebka, G. A. 1959, PhRvL, 3, 439

Reines, F., & Cowan, C. L. 1956, Natur, 178, 446

Voges, W. 1993, AdSpR, 13, 391

Weiss, R. 1972, Quarterly Progress Report, Research Laboratory of Electronics MIT, 105, 54

Wheeler, J. A., & Ford, K. 1998, Geons, Black Holes, and Quantum Foam: A Life in Physics (New York: W.W. Norton, Inc)

Wilson, J. G. 1956, Progress in Cosmic Ray Physics (Amsterdam/New York: North Holland/ Interscience Wiley-Interscience)

Chapter 2

From Dust … The Birth and Death of Stars

The star cluster NGC602 reproduced courtesy of NASA, ESA, and the Hubble Heritage Team.

2.1 Introduction

As early as 2000 BCE Egyptians and Babylonians were using calendars based on the regularity of the motions of the stars and planets. This regularity was also used to steer ships at sea and to locate geographical features on land. The dependability of the stars, coupled with observations of an apparently continuously changing Earth with no visible changes in the stars, led the 4th century BCE philosopher Aristotle

doi:10.1088/2514-3433/ac81b3ch2

(384–322 BCE) to the conclusion that the stars and the Sun never change. The Earth was the center of Aristotle's universe and celestial objects orbited the Earth in purely circular (i.e., "perfect") paths. Within the lunar orbit all was mutable but the stars and the Sun were perfect and unchanging.

Aristotle's influence was lost to the West for centuries after the fall of the Roman Empire, but the compatibility of his ideas with the new religion of Christianity then gave it added impetus. This was enhanced by St. Thomas of Aquinas (1225–1274) in the 13th century. St. Thomas's "Commentaries on Aristotle" had a profound influence on the Church and consequently on Western thought. The joint influence of Aristotle and the Church was so great that in Europe the immutability of astronomical events went unquestioned for over 2000 years.

Many novae, variable stars that brighten up for a short time to a luminosity much greater than that of their normal state, had been recorded by Arabic and Chinese astronomers: the sudden appearance and eventual disappearance of these objects caused the Chinese to refer to them as "guest" stars. European astronomers attributed these and occasional comets to atmospheric events. In 1006 AD astronomers in China, Egypt, Iraq, Japan, and Switzerland recorded the sudden appearance of a star bright enough to be seen in full daylight. The "new star" of 1006 AD was in fact the explosive death of a star: a supernova now known as

Figure 2.1. Left (a): A composite image of SN 1006 depicting X-ray emission detected by the Chandra Observatory in 2003 (blue), optical emission detected with the Curtis Schmidt and DSS telescopes in white and yellow, and radio emission detected with the Very Large Array and Great Britain telescopes in red. (Reproduced courtesy of X-ray: NASA/CXC/Rutgers/G. Cassam-Chena and J. Hughes; Radio: NRAO/AUI/NSF/GBT/VLA/Dyer, Maddalena & Cornwell; Optical: Middlebury College/F. Winkler, NOAO/AURA/NSF/CTIO Schmidt & DSS.) **Right (b):** Close-up of the optical ribbon with the Hubble Space Telescope and the Cerro Tolelo Inter-American Observatory (CTIO): a small part of the expanding supernova remnant dominated by a thin and outwardly moving shock front that heats and ionizes surrounding gas. (Reproduced courtesy of W. Blair [JhU]; NASA, ESA, Hubble Heritage [STScI/AURA].)

SN1006[1] (Figure 2.1(a); Winkler 2014). SN1006 turns out to have been the brightest "star" observed in all of recorded history, remaining visible during daylight for over three months and staying a naked eye object at night for several years after that. The expanding debris cloud from the stellar explosion can still be detected with sensitive telescopes (Figure 2.1(b)). The remnant is now 5.7×10^{14} km (roughly 60 light-years) across and the shockwaves from the original explosion, still expanding, are a source of cosmic rays. However, European astronomers continued to ignore the ever visible evidence of the birth and death of stars until in 1572 the Danish astronomer, Tycho Brahe (1546–1601), observed an event (now called Tycho's supernova) that he could prove took place well beyond the orbit of the Moon. Today we find that the sky is full of highly variable phenomena, with stars continuously being born in dense regions or dying in incredibly energetic explosions.

2.2 Stellar Birth

Stars are formed via the gravitational collapse of clouds of gas and dust. One way the collapse can be initiated is by density enhancements that occur in the spiral arms of galaxies causing the gas and dust to be compressed to the point where gravitational infall can occur (Figure 2.2). The molecular clouds in the spiral arms are

Figure 2.2. A composite HST/Subaru Telescope image of the spiral galaxy, M83, showing the many bright star-forming regions (red) that lie along the spiral arms. (Reproduced courtesy of Subaru Telescope (NAOJ), HST-ESO/Gendler.)

[1] A commonly accepted naming of supernova is to use SN followed by the year, month, and day of their discovery (followed by a letter designation if more than one is detected in a given day).

themselves quite cold (about 10–100 degrees Kelvin) but as matter within them gets condensed it heats up. If sufficient matter collapses to raise the core temperature of the collapsing cloud to 13.6 million degrees Kelvin (the temperature at which hydrogen fuses into helium, releasing energy) a star is born. The energy released by hydrogen fusion is what powers the star. For a star to be stable the energy released by thermonuclear burning needs to balance the gravitational infall causing collapse. It takes a tremendous amount of matter: in a cloud with an initial temperature of 100 degrees Kelvin it would take 10^{57} atoms. This happens to be the number of hydrogen atoms in a star with the mass of our Sun. If the collapsing mass is not large enough to ignite nuclear fusion in its center it becomes a "brown dwarf," too massive to be a planet but not massive enough to be a star. Depending on the amount of matter that collapses, stars are born at masses ranging from about 1/10th to about 100 times the mass of our Sun.

Stars appear to be simple; they form essentially perfect spheres, radiating light from their surfaces uniformly, and a typical solar-mass star, such as our Sun, remains stable for several billion years before any observable changes occur. However, stellar birth is not so simple. Young stars are rarely, if ever, surrounded by spherical envelopes of gas and dust. This is because the clouds within which the stars are born themselves are not uniform; their densities and temperatures vary by orders of magnitude and their shapes are amorphous (Figure 2.3). Also they have magnetic fields that can work to inhibit gravitational collapse. Hence, the clouds do not collapse simply into spheres but produce jets, stellar winds, and possibly planet-forming disks of dust (Figures 2.4–2.6; Burrows et al. 1996; O'Dell 1998).

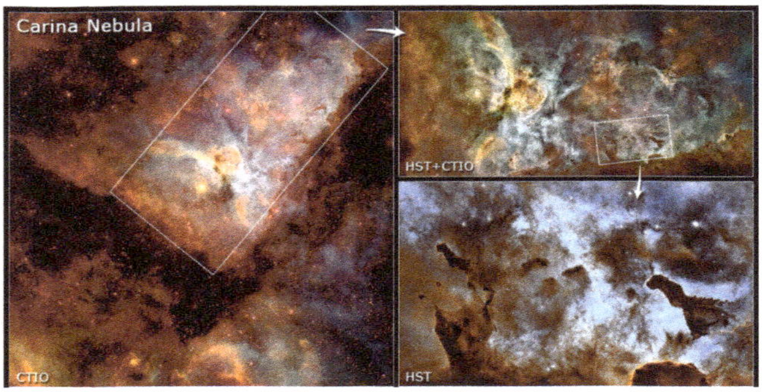

Figure 2.3. Left (a): Cloud formations of the Carina Nebula as imaged by ground-based telescopes at the Cerro Telello International Observatory in 2003. The Nebula is over 300 light-years wide and lies about 7500 light-years from Earth. Colors depict emission from oxygen (blue), hydrogen (green), and sulfur (red). (Reproduced courtesy of NASA, ESA, N. Smith (Univ. California, Berkeley) et al. and The Hubble Heritage Team (STScI/AURA).) **Upper right (b):** A composite of 48 high-resolution images taken by Hubble showing a 60 light-year span. The Hubble image is combined with CTIO filter images to include color. The nebula is composed of knots of molecular gas and dust so thick they have become opaque. Eta Carinae, the bright white ovoid spot toward the left center of the image is the most energetic star in the nebula, and was one of the brightest stars in the sky in the 1830s. **Lower right (c):** A close-up of the Carina Nebula pillars (each roughly one light-year long) composed of Hubble/ACS images taken in light emitted by hydrogen and nitrogen atoms (orange) in 2005 and in light emitted by oxygen atoms (blue) in 2010. (Reproduced courtesy of Nathan Smith (Univ. of Arizona), NASA/ESA, and L. Frattare (STScI).)

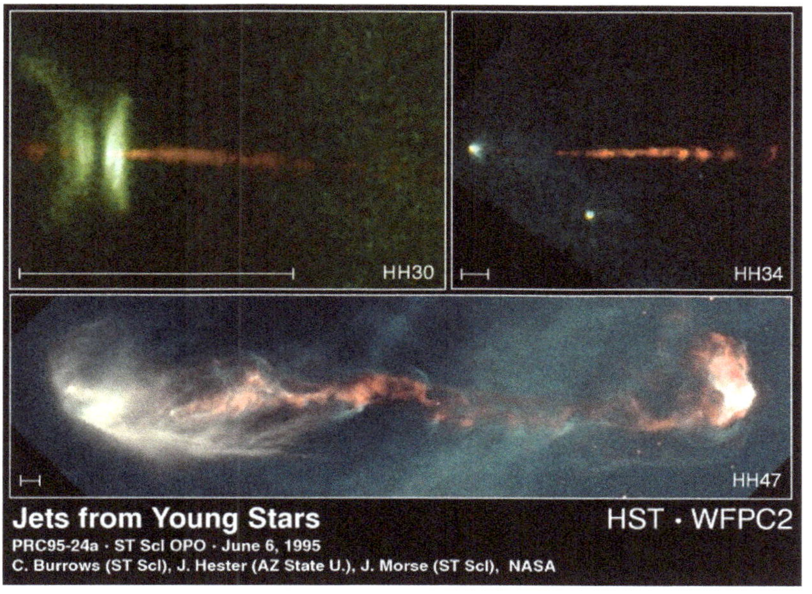

Figure 2.4. Hubble optical views of jets from young stars. **Upper left (a):** The object HH30 shows a disk (in pale green) that is illuminated by a star that is itself hidden. The reddish jet comes from the inner region of the disk, possibly directly from the star itself. **Upper right (b):** the beaded structure from HH34 is produced by clumps of dense gas ejected from the star at speeds of one-half million miles per hour (2.2×10^5 m s^{-1}). **Lower panel (c):** The bright blobs at either end of the three trillion miles (4.8×10^{12} km) long HH47 shows where jet material has slammed into interstellar gas. The very complicated jet pattern might be due to the star wobbling in response to the gravitational pull of a companion star. The scale in the bottom left corner of each picture represents 93 billion miles (15 trillion km), or 1000 times the distance between Earth and the Sun. HH objects are named for their discoverers: astronomers George Herbig and Guillermo Haro. (Reproduced courtesy of NASA/C. Burrows, J. Hester, and J. Morse.)

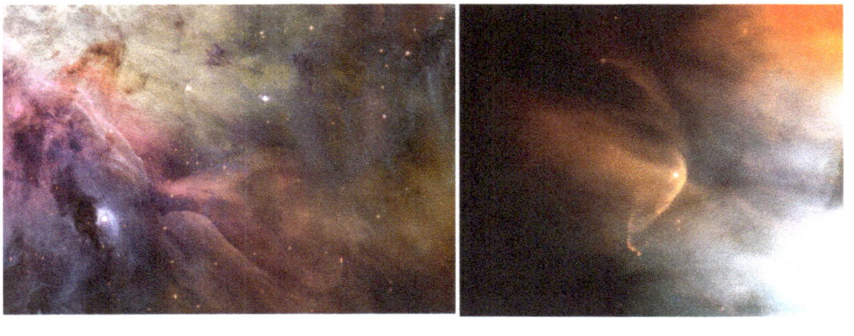

Figure 2.5. Left (a): A Hubble image of clouds and stellar winds from the variable star LL Orionis in the Orion Nebula. Since the material ejected in the stellar wind is faster than the surrounding material it forms a shock front similar to the bow wave of a boat moving in water. The paraboloid arc shape slightly left and up from the center is LL Ori's bow shock. It is roughly half a light-year across. **Right (b):** A close-up of LL Ori's bow shock; a bow shock from another star can be seen in the upper right of this image. (Reproduced courtesy of NASA, ESA, and the Space Telescope Science Institute, and C. R. O'Dell, Vanderbilt University.)

Figure 2.6. Perhaps 80% of young stars are surrounded by dusty clouds of gas, variously known as nebulae, protoplanetary disks, or proplyds. These clouds rapidly flatten into extended rotating disks and disperse on timescales of 1 to 10 million years. In most disks, solids condense out of the gases and begin to assemble rocky or icy bodies. The viscous gobs in this photograph are proplyds surrounding newborn stars in the Orion Nebula. The large proplyd in the lower left is about 1000 times the distance of our Earth from the Sun. (Reproduced courtesy of NASA/C.R. O'Dell/Rice University/ESA, and the Space Telescope Science Institute.)

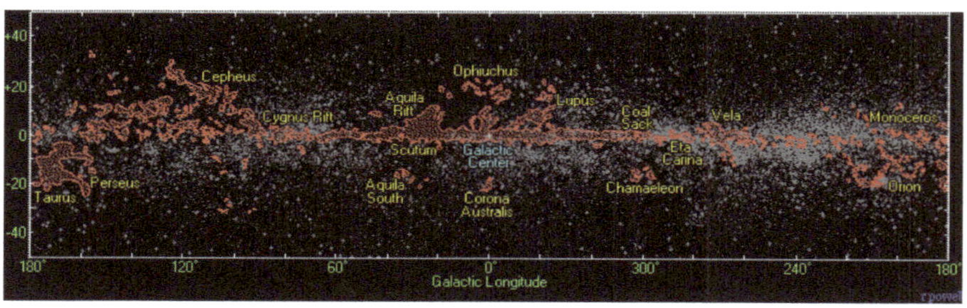

Figure 2.7. A schematic of the Milky Way showing locations of the largest molecular cloud complexes (outlined in red) as identified by observations at millimeter radio wavelengths. The Carina molecular cloud is at Galactic longitudes around 280° and the Orion molecular cloud at Galactic longitudes around 210°. (Reproduced courtesy of Dame, Hartmann, and Thaddeus, Harvard-Smithsonian CfA.)

It is difficult to see starbirth in optical light. The dust in the compressed regions is of roughly the same size as optical wavelengths hence it efficiently absorbs or completely blocks off the optical light. Radio observations are particularly useful for identifying the dense molecular clouds (Figure 2.7; Dame et al. 2001) where star formation occurs, and infrared wavelengths can penetrate the dust and reveal the star-forming regions (Figure 2.8). As angular resolution improved, the jets were seen to emerge along the polar axes of circumstellar disks of gas and dust. The disks surround the stars for the first few million years of their lives dropping matter into

Figure 2.8. Left (a): The Trifid Nebula as imaged in optical light by a 0.9 m telescope at Kitt Peak National Observatory. Red depicts "emission nebulae": a cloud of hydrogen gas activated by intense radiation from newly formed, very massive stars at the center of the nebula. Blue depicts "reflection nebulae": dust reflecting starlight. The dark bands are "dark nebulae": dusty clouds that block the visible light. (Reproduced courtesy of NOAO.) **Right (b):** This infrared image of the Trifid obtained with the Spitzer Space Observatory, shows that the dust bands are composed of bright star-forming regions. Red, green, and blue represent different wavelengths with red the longest and blue the shortest. (Reproduced courtesy of NASA/JPL-Caltech/J. Rho, SSC/Caltech.)

the star and putting energy and mass into the jets. Less dense material above and below the relatively thin disks adds mass to the disks and fuel to the winds. While the collimation (funneling into channels) of the winds into very narrow jets remains to be understood, it is generally believed that mass loss and collimation are essential for starbirth since it provides a means of dispersing excess angular momentum or magnetic energy.

An example of the importance of adding an infrared eye is shown by the images of the Trifid Nebula depicted in Figure 2.8 (Rho et al. 2006). The Trifid Nebula was discovered about 250 years ago by the French astronomer Legentil de La Galaisiere (1725–1792), and was identified as M20 in a catalog of nebulous objects compiled by Charles Messier (1730–1817). The name Trifid, meaning split three ways, arose because that was how it looked in 19th Century telescopes. The dark lanes in the visible show up as bright star-forming regions in the infrared image obtained with Spitzer. A striking example of the different views from optical and infrared wavelength regions is illustrated in Figure 2.9, where the optical image of one of the pillars from the Carina Nebula (Figure 2.3) barely shows the enormous jets that are distinct in the infrared image. An animation comparing the optical and infrared images in Figure 2.9 is shown in Figure 2.10(a).

Figure 2.9. **Upper panel (a):** An optical image of a two light-year wide pillar from the Carina Nebula. The pillar's convoluted outlines are shaped by the winds and radiation of young, hot, massive stars and within it are stars in the process of forming. (Reproduced courtesy of NASA, ESA, and the Hubble SM4 ERO Team.) **Lower panel (b):** In the near-infrared the pillar is dominated by two, narrow, energetic jets (shown in white) blasting outward from a still hidden infant star. Both visible light and near-infrared images were made using the Hubble Space Telescope's newly installed Wide Field Camera 3. (Reproduced courtesy of NASA, ESA, and the Hubble SM4 ERO Team.)

Perhaps the best known star-forming region in our Galaxy, approximately 1500 light-years from our Sun, is within the Orion Nebula (Figure 2.10(b)). Appearing to the naked eye during winter nights as the bright "star" in the sword of Orion, the nebulosity becomes apparent with binoculars. This is our closest massive star formation factory, extending from the center to the edge of the nebula, where hydrogen and other less abundant gases (e.g., nitrogen, oxygen, and sulfur) are heated and ionized by ultraviolet radiation from the star cluster at the center. The gas in the nebula is tenuous, with a highly uneven distribution. Interstellar dust clouds are also present, absorbing light (especially ultraviolet light) so that the gas lying in their shadow glows more dimly than the surrounding areas.

NGC 604 (Figure 2.11) is a site where new stars are being born in a spiral arm of the galaxy M33, it encompasses a region that is almost a hundred time larger than the Orion Nebula star-forming region in our Galaxy (Tüllmann et al. 2008). Though such nebulae are common in galaxies, this one is particularly large (1.4×10^{16} km or 1500 light-years across), nearly as far across as the distance from us to the Orion Nebula! Over 200 hot stars lie in the center of NGC 604 (each about 15 to 16 times the mass of our Sun) heating the surrounding interstellar cloud with ultraviolet light. By studying the physical structure of such giant nebulae, we may determine how clusters of massive stars affect the evolution of the interstellar medium of the galaxy.

Starbirth produces extreme stellar winds that bombard the dusty molecular clouds. As energy from nearby, newly-formed hot young stars pours into the cool dust and gas, large temperature and pressure differences can be created, generating

Figure 2.10. Left (a): An animation comparing the optical and infrared images of one of the pillars from Figure 2.3. (Reproduced courtesy of NASA/Hubble.) Animation available at http://doi.org/10.1088/2514-3433/ac81b3. **Right (b):** A composite of infrared, visible, and ultraviolet observations of the Orion Nebula. Green represents hydrogen and sulfur gas observed with Hubble's ultraviolet and visible-light instruments. The gas has been heated and ionized by intense ultraviolet radiation called polycyclic aromatic hydrocarbons in the cloud. On Earth, polycyclic aromatic hydrocarbons are found on burnt toast and in automobile exhaust. The green specks represent stars that are not as deeply embedded in the nebula. Blue specks are foreground stars. Orange-yellow dots represent young stars revealed in the infrared by the Spitzer Space Telescope. Stellar winds from clusters of newborn stars scattered throughout the cloud formed the ridges and cavities in Orion. The large cavity near the right of the image was most likely carved by winds from four central stars known as the Trapezium. (Reproduced courtesy of NASA/JPL-Caltech/STScI).

Figure 2.10. (Continued.)

winds. The Hubble optical and Spitzer infrared images in Figure 2.12 show examples of this. As may occur in the spectacular phenomena of Earth tornadoes, the large difference in temperature between the hot surfaces of the stars and the cold interior of the clouds, combined with the pressure of starlight, produce strong

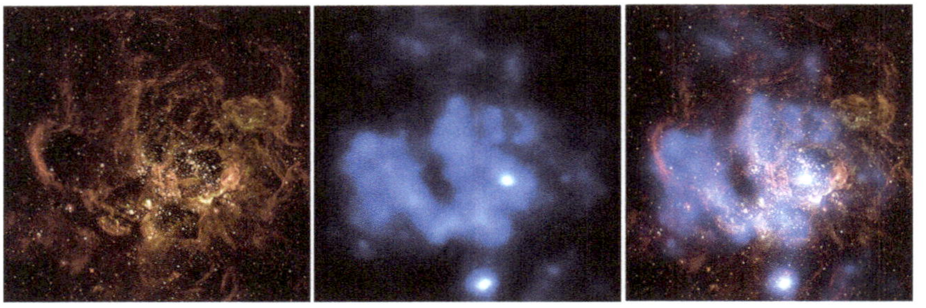

Figure 2.11. **Left (a):** A detailed Hubble optical image of NGC 604 taken in 2003 shows more than 200 stars. The most massive stars in NGC 604 exceed 120 times the mass of our Sun, and their surface temperatures are as hot as 72,000 degrees Fahrenheit (40,000 Kelvin). Ultraviolet radiation floods out from these hot stars, making the surrounding nebular gas fluoresce. (Reproduced courtesy of NASA/AURA/STScI.) **Center (b):** The X-ray emission of NGC 604 is divided by a wall of relatively cool gas. On the western (right) side of the nebula, this X-ray radiation is likely due to the energetic winds from the young, massive stars visible in the optical image. On the eastern side the X-ray filled cavities seem to be older, and are possibly due to past supernova explosions. **Right (c):** Combining the X-ray and optical images shows that the X-ray emission (blue) is confined to the cavities blown in the cooler gas by powerful stellar winds. Note that the brightest spots of X-ray emission (white) have no optical counterparts. (Reproduced courtesy of NASA/CXC/CfA/R. Tuellmann et al.)

Figure 2.12. **Left (a):** Hubble/WFPC image of the Lagoon Nebula, Messier 8, which lies 5000 light-years away in the direction of the constellation Sagittarius. Three narrow-band filters were used with red light indicating ionized sulfur atoms, blue light indicating double ionized oxygen atoms, and green light indicating ionized hydrogen. The color-coded image is the combination of two observations calibrated and combined from the public archive. (Reproduced courtesy of A. Caulet (ST-ECF, ESA), NASA.) **Right (b):** A Spitzer infrared view of HH 49/50. The newly born star responsible for the 100 kilometer per second jet is located just off the top of the picture, while the bright star seen near the tip of the jet is by chance along the line of sight. The nebula glows with infrared light generated as the outflow heats surrounding dust clouds. The color coding shows a trend from red to blue hues at the tornado's tip indicating a systematic increase in emission at shorter wavelengths. HH49/50 is about 450 light-years distant, located in the Chamaeleon I molecular cloud. (Reproduced courtesy of J. Bally (Univ. of Colorado) et al. 2006, JPL-Caltech, NASA.)

horizontal shear to twist the clouds into their tornado-like appearance. Though the spiral shapes suggest the clouds are "twisting," more powerful instruments such as the James Webb Space Telescope (JWST; Chapter 7) are needed to actually measure the wind velocities and to determine the direction of the flows.

2.3 Stellar Death

While stellar birth is best observed at wavelengths longer than optical (such as infrared and radio) because of obscuring dust, stellar death is associated with processes at very high temperatures, so that much of the energy comes out in wavelengths shorter than optical (such as ultraviolet and X-ray) or as energetic particles, such as neutrinos. How a star's life ends depends mostly on its initial mass. The more massive a star, the shorter its lifetime, and the more brilliant its end. Catastrophic supernova explosions identify the death of massive stars, whereas the more common, low-mass stars, such as our Sun, die by ejecting their outer shells in slowly-expanding "planetary nebulae."

2.3.1 Planetary Nebulae: The Death of Sun-like Stars

As they exhaust their thermonuclear fuel, intermediate-mass stars like our Sun expand, becoming cooler and redder, increasing their sizes and energy output tremendously; at this stage they are called red giants. Most of the carbon (the basis of life) and particulate matter (crucial building blocks of solar systems like ours) in the universe are manufactured and dispersed by red giant stars. In their final stage, red giants shed their outer bulk, creating an expanding cloud of gas. The small, hot remnant of the central star heats the gas, causing it to glow. For several thousand years the dying star is surrounded by this glowing gas cloud known as a planetary nebula. Planetary nebulae were so called because over 200 years ago astronomers peering through small telescopes thought they were shaped like planets. However, unlike planets, they did not move with respect to the background stars. Planetaries are among the brightest nebulae known. Unlike stars, which radiate continuously across the spectrum, the light emitted by a planetary nebula is made up of emission at discrete wavelengths of atomic lines, rather like a fluorescent or neon light.

A long-standing puzzle is how planetary nebulae acquire their complex shapes and symmetries, since red giants and the gas/dust clouds surrounding them are mostly spherical. A slowly-expanding stellar wind is likely to be spherically symmetric, but almost all well-resolved planetary nebulae images deviate substantially from sphericity, having rather large-scale amorphous, elliptical, or bipolar shapes (Figure 2.13). Hubble's ability to see very fine structural details (usually blurred beyond recognition in ground-based images) enables us to look for clues to this puzzle.

Several theories suggest that such extreme asymmetry can only happen when a binary companion is present; the more spectacular nebulae involve interactions between a large, aging star and unseen companions, like planets or brown dwarfs. The rotational effects of another body nearby would presumably spin up the aging star, setting up conditions that would produce gaseous jets and other dynamic

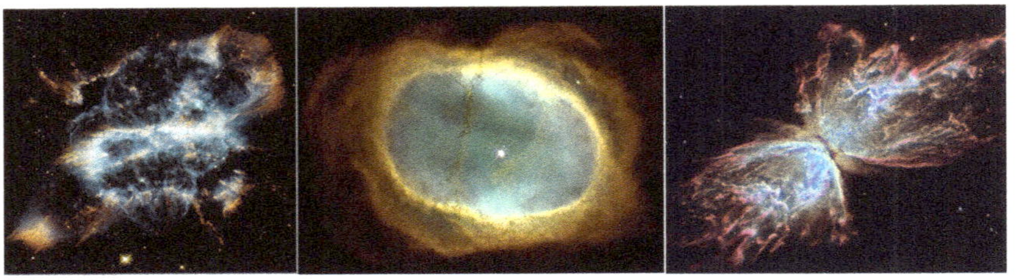

Figure 2.13. **Left (a):** Planetary Nebula NGC 5189 is unusually complex. Shown here in an image combining broad band optical in blue, broad band infrared in yellow, hydrogen in orange, oxygen in cyan, and sulfur in green. The animation shows the nebula in context in its position in the sky. (Reproduced courtesy of NASA, ESA, and the Hubble Heritage Team (STScI/AURA).) Animation available at http://doi.org/10.1088/2514-3433/ac81b3. **Center (b):** NGC 3132 is about half a light-year in diameter and is located 2000 light-years away. The gases emitted by the dimmer of the two stars at the center (a white dwarf) are moving away at speeds around 14 km per second. The brighter companions star will likely lose its envelope at a later date. Blue represents the hottest gas, which is confined to the inner region of the nebula. Red represents the coolest gas, at the outer edge. The dust filaments contain a lot of carbon. (Credit: NASA, ESA, and the Hubble Heritage Team (STScI/AURA). **Right (c):** NGC 6302, the "Butterfly Nebula" is roughly 3800 light-years away in the constellation Scorpius. The star's outer layers, expelled over about 2200 years is more than two light-years across. The star's surface temperature is estimated to be about 400,000 degrees Fahrenheit, making it one of the hottest known stars in our Galaxy. (Reproduced courtesy of NASA, ESA, and the Hubble SM4 ERO Team; reprocessing Francesco Antonucci.)

patterns. Noam Soker of Haifa has argued that more than half of progenitors of planetary nebulae have planets around them; it is the planets that spin up the stars and lead to the wide varieties of morphologies (Soker & Kashi 2012).

Iosif S. Shklovsky (1956) noted similarities between white dwarfs and the hot, central stars of planetary nebulae, correctly deducing that the former evolve from the latter after losing their surrounding nebulae. George Abell and Peter Goldreich (1966) noted the similarity between the expansion velocity of planetary nebulae and the escape velocity (the velocity at which matter can overcome the gravitational pull of the star is called its escape velocity) of red giants. In the mid-70s while studying the planetary nebulae NGC 7027 (Figure 2.14). Sun Kwok and colleagues at York University (1978) found the connection between planetary nebulae and the red giants that they evolve from. Red-giant envelopes are built up over about a million years via a slow stellar wind. If this wind entirely depletes the red giant's atmosphere, a hot core will be exposed. The fast stellar wind that is hypothesized to emerge from this core acts as a snowplow when it encounters the material expelled from the red giant. That material is then swept up into a dense nebula—the planetary—while the much larger red-giant envelope remains beyond. Observational confirmation of this theory followed in 1978: one of the first discoveries made by Sara Heap with the newly launched International Ultraviolet Explorer (IUE) was that fast stellar winds indeed flow from the central stars of planetary nebulae (Heap et al. 1978). About a decade later You-Hua Chu (University of Illinois) and George Jacoby (Kitt Peak National Observatory), using CCD cameras (which allow the measurement of faint objects in the vicinity of bright ones), discovered that many PNs have faint halos, i.e., the original

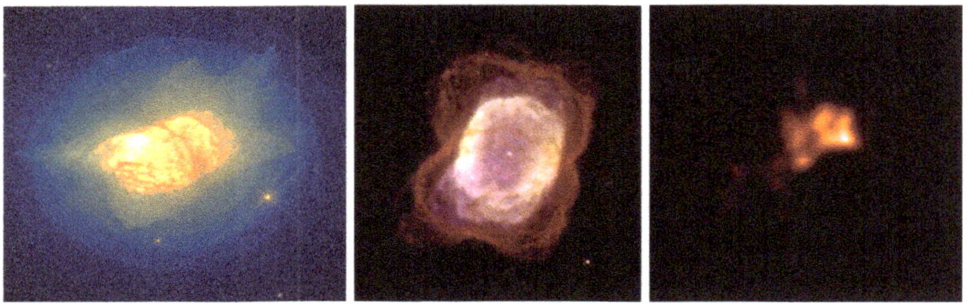

Figure 2.14. **Left (a):** This HST/WFPC2 optical image of the planetary nebula, NGC 7027, shows lanes of red dust clouds through the central bright region and concentric rings (in blue) which show several episodes of mass loss from the original red giant. The white dwarf in the center is barely visible. (Credit: H. Bond (STScI) and NASA). **Center (b):** The HST/NICMOS infrared image of NGC 7027 penetrates the dust to show the white dwarf clearly. Molecular hydrogen is shown in red. (Reproduced courtesy of William B. Latter (SIRTF Science Center/Caltech), NICMOS, HST, NASA.) **Right (c):** The first detection of X-rays from NGC 7027 was made by the Chandra X-ray satellite showing a bubble of 3 million degree Kelvin gas covering an area about a hundred times that of our solar system. The greater brightness to the upper right is due to less obscuring material to block the X-ray emission. (Reproduced courtesy of NASA/RIT/J. Kastner et al.)

Figure 2.15. Chandra and Hubble observations of the bipolar nebula Menzel 3. **Left (a):** An HST image of Mz 3 taken with the Wide Field Planetary Camera shows two sets of bipolar lobes one set spherical and one set conical with some shocked material between them as well filamentary columns. (Reproduced courtesy of NASA.) **Center (b):** The Chandra image shows both a strong central source, diffuse emission confined within both sets of optical lobes, and the shocked regions between the two lobes. (Reproduced courtesy of NASA, Chandra, HST, and J. Kastner (RIT).) **Right (c):** The combined image shows clearly that the X-rays are confined within the optical lobes. (Reproduced courtesy of NASA/Chandra.)

red-giant envelope that is difficult to detect because of its much lower density (Figures 2.14(a) and 2.16(b); Chu et al. 1987). Early in 2001 the Chandra observatory detected X-ray emission from planetary nebulae for the first time (Figure 2.14(c); Kastner et al. 2001). The detection lent further support to the snowplow theory: impact of the fast stellar wind on the earlier ejected material causes the matter to heat up to the several million degrees required for it to glow in X-rays (Figure 2.16). The X-rays are confined within the earlier ejected halos (Figure 2.16(b)).

In the bipolar planetary nebula Menzel 3, Chandra observations showed both a compact source at the center as well as jet like protrusions along the polar axis of the optical nebula (Figure 2.15; Kastner et al. 2003). The observed combination of an X-ray-emitting point source and possible X-ray jet(s) is consistent with models in which accretion disks and, potentially, magnetic fields, shape bipolar planetary nebulae via the generation of fast, collimated outflows.

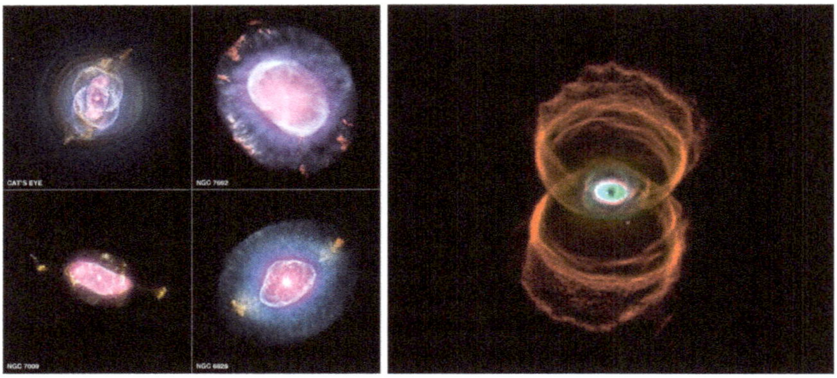

Figure 2.16. Left (a): Hubble (red, green, and blue) and Chandra (red) images of four planetary nebulae. The signature of periodic mass loss from the original star is clearly seen as expanding rings in the Cat's Eye planetary nebula (upper left panel). The diffuse X-ray emission seen with Chandra is caused by shockwaves as a wind from the hot remnant of the star collides with the ejected atmosphere and is confined within the earlier ejected material seen in optical (red, green, and blue). (Reproduced courtesy of X-ray: NASA/CXC/RIT/J. Kastner et al.; Optical: NASA/STScI.) **Right (b):** The planetary nebula MyCn18 is located about 8000 light-years away. In this HST/WFPC2 image hydrogen gas is represented by green, oxygen by blue, and nitrogen by red. A pair of intersecting elliptical rings at the center appear to form a smaller version of the large "hourglass" at right angles. (Reproduced courtesy of R. Sahai and J. Trauger (JPL), WFPC2, HST, NASA.)

In bipolar planetary nebulae the central star itself is often obscured within a ring of dust. The ring constricts the outflow of material from the star creating the bipolar shape. The central star in the planetary nebula M2-9, is known to be one of a pair which orbit one another at very close distances. In such cases the gravity of one star pulls weakly bound gas from the surface of the other and produces a thin, dense disk which surrounds both stars. Models of the type that are used to design jet engines (hydrodynamics) show that such a disk can successfully account for the jet-exhaust-like appearance of M2-9. The high-speed wind from one of the stars rams into the surrounding disk, which serves as a nozzle. The wind is deflected in a perpendicular direction and forms the pair of jets that we see in the nebula's image. This is much the same process that takes place in a jet engine: the burning and expanding gases are deflected by the engine walls through a nozzle to form long, collimated jets of hot air at high speeds.

In ground-based images, the young planetary nebula MyCn18 appears as a pair of large outer rings with a smaller central one, but the Hubble image (Figure 2.16(b)) shows fine details that shed light on the poorly understood ejection of stellar matter which accompanies the slow death of Sun-like stars. According to one theory for the formation of planetary nebulae, the hourglass shape is produced by the expansion of a fast stellar wind within a slowly-expanding cloud that is more dense near its equator than near its poles. What appears as a bright elliptical ring in the center, and at first sight might be mistaken for an equatorially dense region, on closer inspection appears to be an ovoid structure with a symmetry axis dramatically different from that of the larger hourglass. The hot star which has been thought to eject and illuminate the nebula, and therefore expected to lie at its center of symmetry, is clearly off center. Hubble has also revealed other features in MyCn18 that are completely new and unexpected. For example, there is a pair of intersecting elliptical rings in the central region that appear to be the rims of a smaller hourglass. There are the intricate patterns of the etchings on the hourglass walls. The arc-like etchings could be the remnants of discrete shells ejected from the star when it was younger, flow instabilities, or the result of the action of a narrow beam of matter impinging on the hourglass walls. An unseen companion star and accompanying gravitational effects may well be necessary in a theory to explain the structure of MyCn18.

Ring-like planetary nebulae may in fact be bipolar nebulae but seen through one end of the cylinder rather than across. NGC 3132 clearly shows two stars near the center of the nebula, a bright white one, and an adjacent, fainter companion to its upper right (Figure 2.13(b)). (A third, unrelated star lies near the edge of the nebula.) The faint partner is actually the star that has ejected the nebula and is now an exposed core. The brighter star is in an earlier stage of stellar evolution, but in the future it will probably eject its own planetary nebula. The Hubble image also reveals a host of filaments, including one long one that resembles a waistband, made out of dust particles that have condensed out of the expanding gases. The dust particles are rich in elements such as carbon. Eons from now, these particles may be incorporated into new stars and planets when they form from interstellar gas and dust. Stunning images of the ring-like nebula known as the Helix (Figure 2.17) show hundreds of

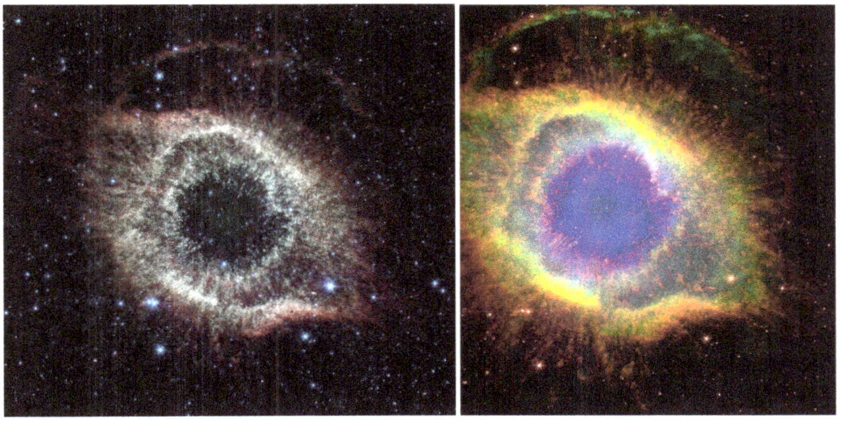

Figure 2.17. Left (a): Emission in this mid-infrared Spitzer Space Telescope image of the Helix comes mostly from the nebula's molecular hydrogen gas. The gas is clumpy, forming thousands of comet-shaped knots each spanning about twice the size of our solar system. Bluer, more energetic radiation is seen to come from the heads with redder emission from the tails, suggesting that they are more shielded from the central star's winds and intense ultraviolet radiation. The nebula itself is about 2.5 light-years across. (Reproduced courtesy of NASA.). The animation shows the same object in optical with Hubble; near, far, and mid-infrared with Spitzer; and ultraviolet with Galex.) Animation available at http://doi.org/10.1088/2514-3433/ac81b3. **Right (b):** A composite of infrared data from Spitzer and visible-light data from NASA's Hubble Space Telescope. (Reproduced courtesy of NASA/JPL-Caltech/J. Hora (CfA), W. Latter (NASA/Herschel, and C. R. O'Dell (Vanderbilt).)

comet-shaped knots formed by material falling back onto the central region. Our own Sun may eject a similar planetary nebula some 6 billion years from now.

2.3.2 Supernovae: The Deaths of Massive Stars

Supernovae are stars that explode far more violently than ordinary novae: the increase in brightness at optical wavelengths can reach a billion (10^9) times the initial brightness of the star whereas ordinary novae only brighten by a factor of about 100,000 (10^5). The brightest supernovae observed can be a hundred times brighter than the galaxy in which they are found. Adding up emission in optical light alone gives a total energy for the supernova explosion of a few times 10^{48} erg. However, supernovae emit at all wavelengths, and integrating emission from radio to X-ray gives another factor of one thousand, so the total energy dissipated by a supernova in a year is of the order of 10^{51} erg—an amount that would take the Sun ten billion (10^{10}) years to radiate! Even more energy is released in the form of neutrinos: up to 10^{53} erg. This energy is a large portion of the total thermonuclear energy potentially present in the star; thus while a nova may flare again and again losing only a small portion of its energy at each flare up, a supernova can explode only once.

In the 19th century Charles Messier recorded a fuzzy ball of light toward the constellation Taurus, which looked similar to a comet, but did not move across the sky and recorded the nebula as the first object in his catalog of things that should not be mistaken for comets. M1, now known as the Crab Nebula (because of its crab-like appearance in low resolution optical observations) consists of the material ejected in a supernova explosion, which has been spread over a volume approximately 9.5×10^{13} km (about 10 light-years) in diameter. It is still expanding at the very high velocity of about 1800 km s^{-1}. By comparing pictures taken several years apart and extrapolating backwards in time it was found that the filaments first started expanding away from the center around 1040–1070 AD. This agrees well with the supernova explosion that was recorded by Arabic, Chinese, and European astronomers in 1054 AD.

In 1948, the Crab Nebula was identified as a strong source of radio radiation (Figure 2.18(a); Bolton & Stanley 1949). With some exceptions, most supernovae remnants are *radio sources*, where the emission is caused by charged particles accelerated along magnetic field lines (synchrotron radiation). The radio emission indicates that strong magnetic fields and charged particles are present. The strong fields are generated by compression of existing weaker fields and by turbulent processes. The particles are also accelerated by these processes and within shocks. Thus the interfaces between regions are expected to be bright in radio waves. In 1968, a pulsating radio source, the Crab Pulsar (also cataloged as NP0532) was discovered in M1. It has now been established that this pulsar is a rapidly rotating neutron star: it rotates about 30 times per second!

The Crab Nebula in the visible spectrum shows two distinct features: a reddish web of filaments at the outer edges of the nebula and a bluish core. Hubble observations of the Crab's inner regions (Figure 2.18(b)) show a complex of sharp knots and wisp-like features that display dynamic variations over a relatively short

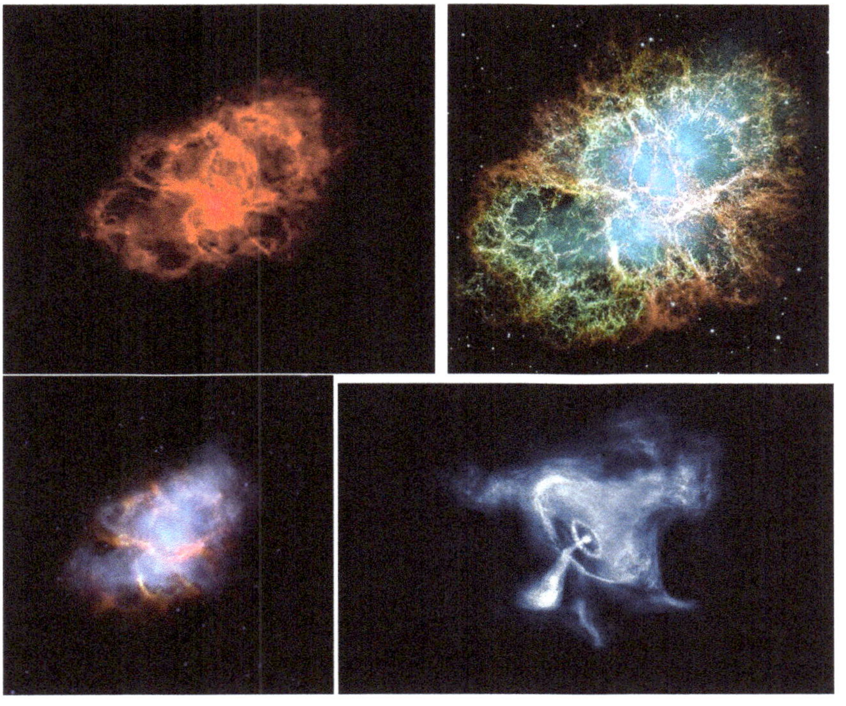

Figure 2.18. **Upper left (a):** A radio image of the Crab Nebula made with the Very Large Array. The radio emission comes from the highly energetic wind from the pulsar. (Reproduced courtesy of NRAO/AUI/NSF.) **Upper right (b):** Hubble optical view of the Crab showing a complex of filaments that is not entirely understood. (Credit: NASA.) **Lower left (c):** In the Spitzer Space Telescope infrared view of the Crab the blue-white region traces the cloud of energetic electrons trapped within the star's magnetic field, emitting so-called "synchrotron" radiation. Data from the Infrared Array Camera (IRAC) and Multiband Imaging Photometer (MIPS) at 3.6 (blue), 8.0 (green), 24 (red) microns. (Reproduced courtesy of NASA.) **Lower Right (d):** An X-ray image of the Crab taken with Chandra shows rings of emission traceable down to the neutron star pulsar at its center. The animation shows the X-ray pulses. (Reproduced courtesy of NASA.) Animation available at http://doi.org/10.1088/2514-3433/ac81b3.

period (months). The core of the nebula is blue from electrons within the nebula being deflected and accelerated by the magnetic field of the central neutron star. The radiation appears blue because this process emits more light in the shorter (bluer) wavelength portion of the visible spectrum than in the longer (redder) wavelength portion.

Detailed infrared observations of the Crab obtained with the Spitzer satellite show synchrotron radiation emitted by energetic electrons trapped within the star's magnetic field (Figure 2.18(c)). The filamentary structure well known from optical images appears clearly in the infrared image confirming the presence of hot gases whose origin is still a mystery. Surprisingly, emission from the dust that is expected to surround the nebula is not detected (Figure 2.19).

Figure 2.19. A composite image of the Crab Nebula using radio data from the VLA in red, infrared data from the SST in yellow, optical data from HST in green, ultraviolet data from XMM in blue, X-rays data from Chandra in magenta. (Reproduced courtesy of NASA—X-ray: CXC, J. Hester (ASU) et al.; Optical: ESA, J. Hester and A. Loll (ASU); Infrared: JPL-Caltech, R. Gehrz (U. Minn).)

Supernova explosions are the most energetic stellar events known and can ultimately be responsible for heating hundreds of solar masses of material to millions of degrees—X-ray emitting temperatures. Not surprisingly they were the *only* class of objects predicted (by Philip Morrison (1915–2005); Morrison 1957) to be X-ray emitters before cosmic X-ray sources were discovered in 1962. X-rays from the Crab were first detected in 1964 using a high-altitude rocket (Giacconi & Gursky 1965); the energy emitted in X-rays by the Crab Nebula is about 100 times what it emits in visible light. The Crab Nebula appears smaller and more condensed in X-rays because the electrons that are primarily responsible for the X-ray emission exist only near the central pulsar. The central star is seen to pulse in X-rays, just like it does at radio and optical wavelengths.

The filaments surrounding the edges of the nebula are what is left of the original outer layers of the star. The red color comes from emission of hydrogen. Blown off the star by the supernova, the filaments are still expanding outward into space, away from the central star. The Crab Nebula in the ultraviolet shows a nebula which is slightly larger than that seen in X-rays. This reveals that cooler electrons (responsible for the UV emission) extend out beyond the hot electrons near the central pulsar. This supports the theory that the central pulsar is responsible for energizing the electrons.

There are only a half-dozen recorded historical supernovae—all were observed before the first astronomical use of the telescope by Galileo (1564–1642). During 1937–1941 a search led by Fred Zwicky led to the discovery of 19 extra-galactic supernovae (Zwicky 1942). Today astronomers using a variety of telescopes find from 15 to 20 supernova events a year—but since 1604 none have been seen in our own Galaxy. Today radio surveys are the major source of "new" SNR detections, distinguishing them from H II regions such as NGC 604 (Figure 2.11) by the relatively flat shape of their radio spectra. SNR shockwaves may trigger the collapse of clouds into new stars; hence, while a supernova event marks the end of the life of one star, it may stimulate the birth of the subsequent generation. Thus the observed close proximity of supernova remnants, the signature of stellar death, and H II regions, the signature of stellar birth. As of 2014, D. A. Green of the Cambridge Astrophysics group cataloged 294 supernova remnants (Green 2014): of these 264 are detected in radio, about 40% are detected in X-rays, and only about 30% in the optical. The low number of X-ray remnants is due to the absorption of X-rays by the interstellar medium (ISM) which limits detection. Optical identifications are difficult because other emission nebulae may resemble remnants in appearance; observations at multiple wavelengths are necessary to confirm supernova-remnant nature.

Early classification of supernova explosions based on optical emission alone suggested two main types, logically called Type I and Type II. The optical light curves of Type I Supernovae show a rapid rise during a few days to maximum brightness near which they may stay for about one or two weeks, followed by a gradual decline in brightness by as much as a factor of a million within a few years. The gradual decline forms an exponential decay curve with a half-life of 55 days, which matches the radioactive decay of Cobalt-60, implying the presence of Cobalt-60 in the remnant. The optical light curves of Type II Supernovae rise more slowly to a maximum that can last for several months, and they are generally less luminous than those of Type I. The Type II maximum is broader, and the decay occurs on less regular timescales; a great deal of individuality is observed, implying a marked variation in the progenitor stars or explosion characteristics. Unlike the Type I where the optical spectra show no hydrogen lines (even though hydrogen is by far the most common element in the universe) the optical spectra of Type II show strong, broad hydrogen lines suggesting that the explosion occurred in young stars which had hydrogen-rich outer envelopes (Figure 2.20).

The most likely mechanism for Type II explosions is core collapse of massive stars (greater than 8 times the mass of our Sun). The subsequent explosion and remnant provide us with the opportunity to observe directly the interiors of stars far more massive than our Sun. Calculations show that massive stars form a dense iron core. After sufficient mass is built up in the core (1.5 Solar masses) the central temperature rises above 3×10^9 Kelvin. The pressure due to thermonuclear reactions that normally supports the outer layers is no longer sufficient, and the star collapses. 10^{53} erg of energy are released, mostly as neutrinos. 10^{51} erg are turned into bulk motion—kinetic energy for heating surroundings. Such an explosion may leave a neutron star, a black hole, or nothing. It is only with multiwavelength (and multi-particle!) observations that the full energy release from a supernova explosion can be determined.

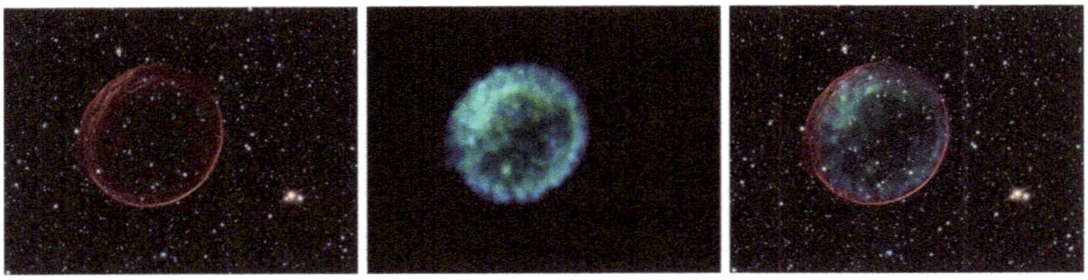

Figure 2.20. Left (a): The supernova remnant NR 0509-67.5 is 160,000 light-years away in the Large Magellanic Cloud. The shell of gas, which is expanding at more than 5000 km/s, is imaged in hydrogen gas (red) by HST/ACS and the background stars were observed with HST/WFC3 in November 2010. (Reproduced courtesy of NASA, ESA, and the Hubble Heritage Team (STScI/AURA).) **Center (b):** X-ray observations with Chandra (blue and green) taken in 2000 and 2007 show material in the center of the remnant that has been heated to millions of degrees. (Reproduced courtesy of NASA/CXC/SAO/J. Hughes et al.) **Right (c):** The combined X-ray and optical image. (Reproduced courtesy of NASA.)

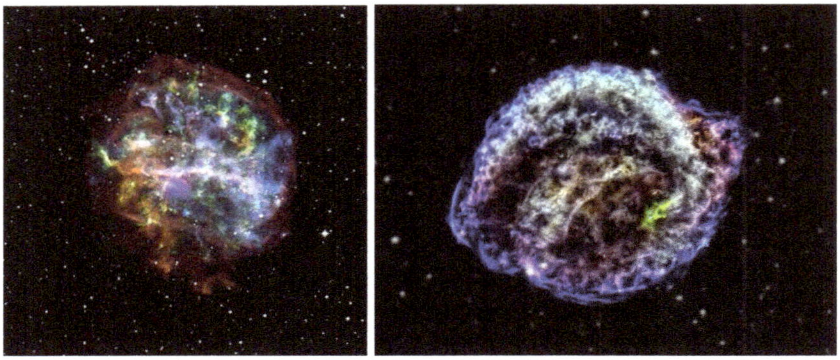

Figure 2.21. Left (a): A remnant left from the explosion of a massive star, this image of G292.0+1.8 shows X-ray emission obtained with the Chandra X-ray Observatory in colors and optical emission obtained through the Digital Sky Survey (DSS) in white. (Reproduced courtesy of X-ray: NASA/CXC/Penn State/S. Park et al.; Optical: Pal.Obs. DSS.) **Right (b):** Chandra image of Kepler's supernova remnant established it as a type 1a. (Reproduced courtesy of NASA/CXC/NCSU/S. Reynolds.)

A possible mechanism for Type I explosions is detonation or deflagration of a carbon-oxygen white dwarf. A white dwarf accreting mass from a companion star approaches a critical core mass of about 1.4 Solar masses. There is no extended hydrogen envelope. Carbon-burning energy output exceeds the neutrino losses, and the star explodes. In deflagration approximately half of the mass is converted to Fe and the rest to intermediate burning elements. In detonation most of the mass is converted to Fe. Later on, the decay of nickel-56 powers the observed emission (Figure 2.21).

While large, massive stars end their furious lives in spectacular supernova explosions—small, low-mass stars may encounter a similar fate. Instead of simply cooling off and quietly fading away, some white dwarf stars in binary star systems are thought to draw enough mass from their companions to become unstable, triggering a nuclear detonation. The resulting standard candle stellar explosion is classified as a Type Ia supernova, and perhaps the best example yet of the aftermath

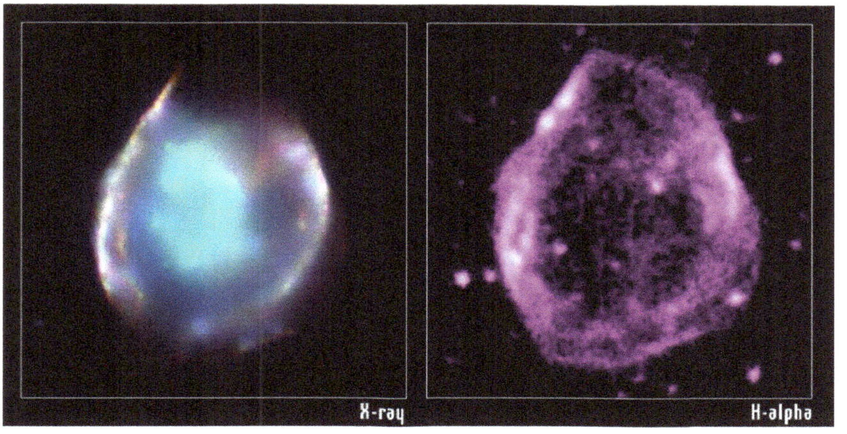

Figure 2.22. **Left (a):** Chandra X-ray image of the supernova remnant DEM L71 shows hot ten million degree iron and silicon gas (aqua) surrounded by a blast wave (white). (Reproduced courtesy of NASA/Chandra/CXC/Rutgers/J. Hughes et al.) **Right (b):** The optical image taken with the 1.5 m CTIO telescope shows only the outer blast wave. (Reproduced courtesy of Rutgers Fabry–Perot.)

is this expanding cloud of shocked stellar debris, DEM L71 (Figure 2.22, Hughes et al. 2003), in the nearby Large Magellanic Cloud. The sharp false-color X-ray image from the orbiting Chandra Observatory shows the predicted bright edges of the outer blast wave shock region and the X-ray glow of an inner region of reverse shock-heated gas. Based on the Chandra data, estimates for the composition and total mass of expanding gas strongly suggest that this is all that remains of a white dwarf star. Light from this small star's self-destructive explosion would have first reached Earth several thousand years ago.

The closest supernova observed since 1604 (Kepler's), was an explosion that took place in our Galaxy's nearest satellite galaxy, the Large Magellanic Cloud. Supernova 1987A was discovered by Ian Shelton and Oscar Duhalde at the Las Campanas Observatory in Chile on 1987 February 24, and within the same 24 hours independently by Albert Jones in New Zealand (Kunkel et al. 1987). Shelton had stepped out to check the sky and noticed a bright star that should not have been there. It inspired him to go to graduate school in the fall of the same year. Fortunately SN1987A occurred at a time when an unprecedented number of space and ground-based observatories were available, making it one of the most extensively observed astrophysical objects in multiple wavelengths. As of 2021, the Astrophysics Data System lists over 1500 papers related to SN 1987A.

To date SN 1987A is the *only* supernova explosion for which information was obtained outside the electromagnetic spectrum: neutrinos from the initial blast were detected by Kamiokande II in Japan, Baksan in the Soviet Union, and the Irvine–Michigan–Brookhaven in the US (Hirata et al. 1987; Alexeyet et al. 1988; Haines et al. 1988). With the significantly improved neutrino detectors today any future Galactic supernova could provide significant insight on neutrino oscillations and supernova physics.

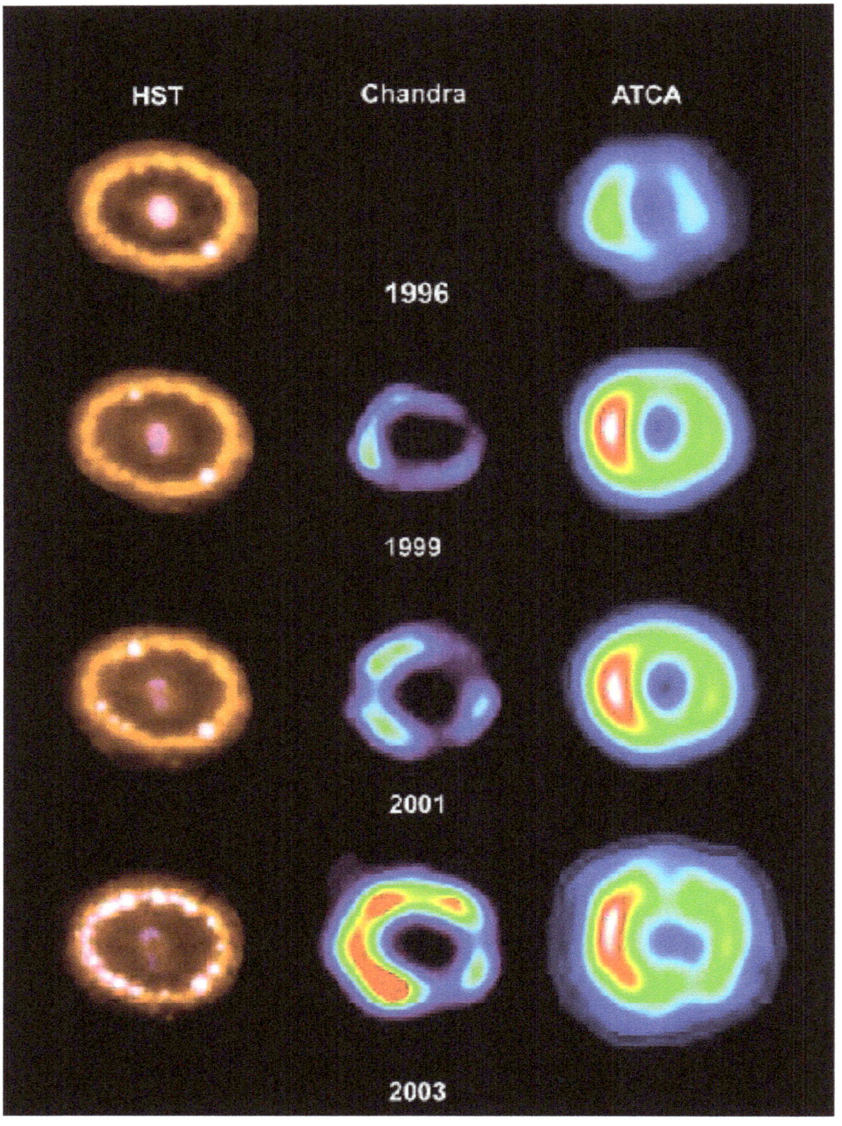

Figure 2.23. The expansion of the SNR from SN 1987A at visible (HST), X-ray (Chandra; launched in 1988) and radio (ATCA) wavebands. The brightening of the outer ring was predicted well before the event by Richard McCray. (Reproduced courtesy of NASA/R. McCray (University of Colorado), D. Burrows and S. Park (Pennsylvania State University), and R. Manchester (Australia Telescope National Facility).)

A paper predicting spectacular X-ray, ultraviolet, and infrared displays proved to be prescient (McCray et al. 1987; Figure 2.23). The striking Hubble picture from 1994 in Figure 2.24 was constructed by the Hubble Heritage Team using observations taken over several years in five filters (B, V, R, ionized oxygen, and hydrogen-alpha) with the Wide Field Planetary Camera 2. The smaller "center" ring of the trio

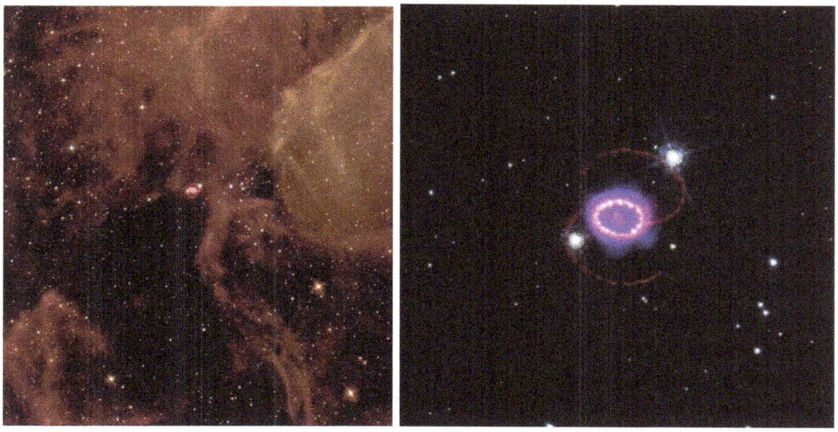

Figure 2.24. **Left (a):** A composite of pictures of SN1987A taken with the Wide Field and Planetary Camera 2 on NASA's Hubble Space Telescope in September 1994, February 1996, and July 1997 shows the remnant and its surroundings. The many bright blue stars nearby the supernova are massive stars, each more than six times the mass than our Sun. They are members of the same generation of stars as the star that went supernova about 12 million years ago. The bright gas clouds shows that star formation is still continuing. (Reproduced courtesy of The Hubble Heritage Team (AURA/STScI/NASA).) **Right (b):** In this image taken in 2011 the brightening of the equatorial ring is obvious. The two bright stars are just in the field of view and are not associated with the supernova. (Reproduced courtesy of NASA/STScI/CfA/P. Challis.)

was seen earlier. The larger pair of outer rings were also seen in ground-based images, but an interpretation was not possible until the higher resolution Hubble observations. Although the rings appear to intersect this is probably an illusion caused by their inclination to our point of view; they are likely in three different planes. The small bright ring lies in a plane containing the supernova; the two larger rings lie in front and behind it. The dim area between the blobs may be related to the equatorial belt of material seen around the supernova that existed before the star exploded. The ring was illuminated by the supernova in 1987 during the explosion and has been slowly fading since then.

The explosion of the supernova debris appears to be perpendicular to the plane of the inner ring. This suggests that whatever properties that the pre-supernova star had, such as rotation or the existence of a companion star, that is responsible for the formation of the inner ring, may also have influenced the dynamics of the explosion. The debris collided with the inner ring by the year 2002 as predicted lighting up all of the dark nebulosity surrounding the supernova (McCray 2003).

Identification of many massive elements within supernovae remnants led to the conclusion that they were formed during a supernova explosion. Twelve years after Baade demonstrated the Type I nature of Brahe's and Kepler's supernovae speculations that some of the observed atomic nuclei in stars may have been formed in explosive situations reached a culmination. In 1957, A. G. W. Cameron and Burbridge, Burbridge, Fowler, and Hoyle (in a paper that astronomers affectionately refer to as B^2FH; Cameron 1957; Burbridge et al. 1957) suggested that nearly *all* the nuclear reaction mechanisms needed for nucleosynthesis were provided during a supernova explosion. Some atomic species are created only in explosive nucleosynthesis. The supernova explosions scatter this material back into space where it is used to create new generations of stars. This is the mechanism by which the gas and dust that formed our solar system became enriched with the elements that sustain life on this planet. X-ray spectroscopy facilitates measurement of the abundances of O, Ne, Mg, Si, S, and Fe in shock-heated material, and thus allows a determination of how enriched it has become from supernova nucleosynthesis products, or in evolved (i.e., older) remnants a determination of the composition of the gaseous interstellar medium that has been swept up by the expanding remnant. Hence Carl Sagan's oft-quoted remark: "We are made of star stuff" (Sagan 1973).

References

Abell, G. O., & Goldreich, P. 1966, PASP, 78, 232

Alexeyev, E. N., Alexeyeva, L. N., Krivosheina, I. V., & Volchenko, V. I. 1988, PhLB, 205, 209

Bally, J., Walawender, J., Luhman, K. L., & Fazio, G. 2006, AJ, 132, 1923

Bolton, J. G., & Stanley, G. J. 1949, AuSRA, 2, 139

Burbidge, E. M., Burbidge, G. R., Fowler, W. A., & Hoyle, F. 1957, RvMP, 29, 547

Burrows, C. J., Stapelfeldt, K. R., Watson, A. M., et al. 1996, ApJ, 473, 437

Cameron, A. G. W. 1957, PASP, 69, 201

Chu, Y-H., Jacoby, G. H., & Arendt, R. 1987, ApJS, 376, 150

Dame, T. M., Hartmann, D., & Thaddeus, P. 2001, ApJ, 547, 792

Giacconi, R., & Gursky, H. 1965, SSRv, 4, 151

Green, D. A. 2014, BASI, 42, 47

Haines, T., Bratton, C. B., Casper, D., et al. 1988, NIMPA, 264, 28

Heap, S., Boggess, A., Holm, A., et al. 1978, Natur, 275, 385

Hirata, K., Kalita, T., Koshiba, M., et al. 1987, PhRvL, 58, 1490

Hughes, J. P., Ghavamian, P., Rakowski, C. E., & Slane, P. O. 2003, ApJ, 582, L95

Kastner, J. H., Balick, B., Blackman, E. G., et al. 2003, ApJ, 591, L37

Kastner, J., Vrtilek, S. D., & Soker, N. 2001, ApJ, 550, 189

Kunkel, W., Madore, B., Shelton, I., et al. 1987, IAU Circ., 4316, 1

Kwok, S. 1978, ApJ, 219, L125

McCray, R. 2003, LNP, 598, 219

McCray, R., Shull, J. M., & Sutherland, P. 1987, ApJ, 317, L73

Morrison, P. 1957, RvMP, 29, 235

O'Dell, C. R. 1998, AJ, 115, 263

Rho, J., Reach, W. T., Lefloch, B., & Fazio, G. G. 2006, ApJ, 643, 965

Sagan, C. 1973, The Cosmic Connection. An Extraterrestrial Perspective (Garden City, NY: Anchor Press, Doubleday)

Shklovsky, I. S. 1956, AZh, 33, 315

Soker, N., & Kashi, A. 2012, ApJ, 746, 100

Tüllmann, R., Gaetz, T. J., Plucinsky, P. P., et al. 2008, ApJ, 685, 919

Winkler, P. F., Williams, B. J., Reynolds, S. P., et al. 2014, ApJ, 781, 65

Zwicky, F. 1942, ApJ, 96, 28

Chapter 3

To Dust … The Birth and Death of Planets

Titan and the rings of Saturn. Reproduced courtesy of NASA/JPL Space Science Institute.

3.1 Introduction

The 6000 or so stars visible to the naked eye are indeed mostly unchanging, and the scarcity of naked eye supernovae (about a half-dozen events in recorded history) was such that they could to some extent be ignored. But in the 17th century, when Galileo Galilei (1564–1642) first turned a telescope to the stars, opening the first "new eye" on a crowded, dazzling universe, he started a scientific revolution that continues to this day (Figure 3.1). The stars were too numerous to count, Venus orbited the Sun, the Moon was cratered, the Sun was NOT perfect and unchanging, and there were planets that orbited other planets!

From being the center of the universe, and being orbited by several thousand celestial bodies, we have become but a speck orbiting one Sun among 100 billion, trillion (10^{23}) suns: there are 10^{11} stars in our Galaxy and 10^{12} galaxies in the known

doi:10.1088/2514-3433/ac81b3ch3

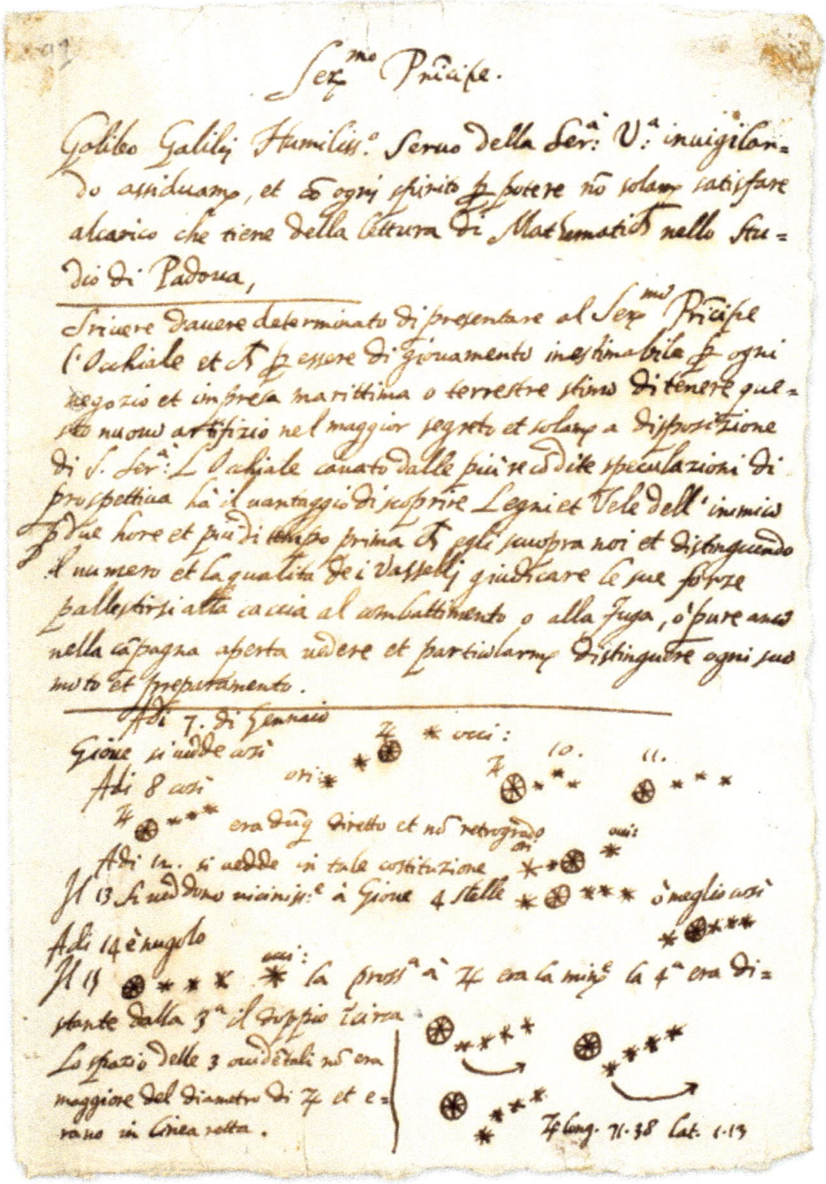

Ser.mo Principe.

Galileo Galilei Humiliss.o Servo della Ser.tà V.a invigilando assiduam.te, et con ogni spirito per potere non solam.te satisfare al carico che tiene della lettura di Matematiche nello Studio di Padova,

Havere determinato di presentare al Ser.mo Principe l'Occhiale et per essere di giovamento inestimabile per ogni negozio et impresa marittima o terrestre stimo di tenere questo nuovo artifizio nel maggior segreto et solam.te a disposizione di S. Ser.tà L'Occhiale cavato dalle più recondite speculazioni di prospettiva ha il vantaggio di scoprire Legni et Vele dell'inimico per due hore et più di tempo prima che egli scuopra noi et distinguendo il numero et la qualità dei Vasselli giudicare le sue forze pallestirsi alla caccia al combattimento o alla fuga, o pure ancora nella campagna aperta vedere et particolarm.te distinguere ogni suo moto et preparamento.

Adi 7. di Gennaio Giove si vedde cosi … cosi:

Adi 8 cosi … era dunq. diretto et non retrogrado

Adi 12. si vedde in tale costituzione

Il 13 si veddono vicinis.e à Giove 4 stelle … ò meglio cosi

Adi 14 è nugolo

Il 15 … la pross.a à Giove era la min.e la 4.a era distante dalla 3.a il doppio in circa

Lo spatio delle 3 occidentali non era maggiore del diametro di Giove et erano in linea retta.

Giove long. 71.38 Lat. 1.13

Figure 3.1. The draft of a letter by Galileo noting his construction of a telescope from a description of one built by the Dutch optician Jan Lippershey. The actual letter, sent by Galileo to the Doge of Venice on 1609 August 24, suggests the usefulness of such a device for war. Below the letter draft is a sketch of the four moons of Jupiter detected by Galileo using his new telescope in January 1610. Paper was precious. (Public domain image of page is at the University of Michigan Harlan Hatcher Graduate Library's Special Collections. The original of the letter sent to the Doge is in the State Archives in Venice.)

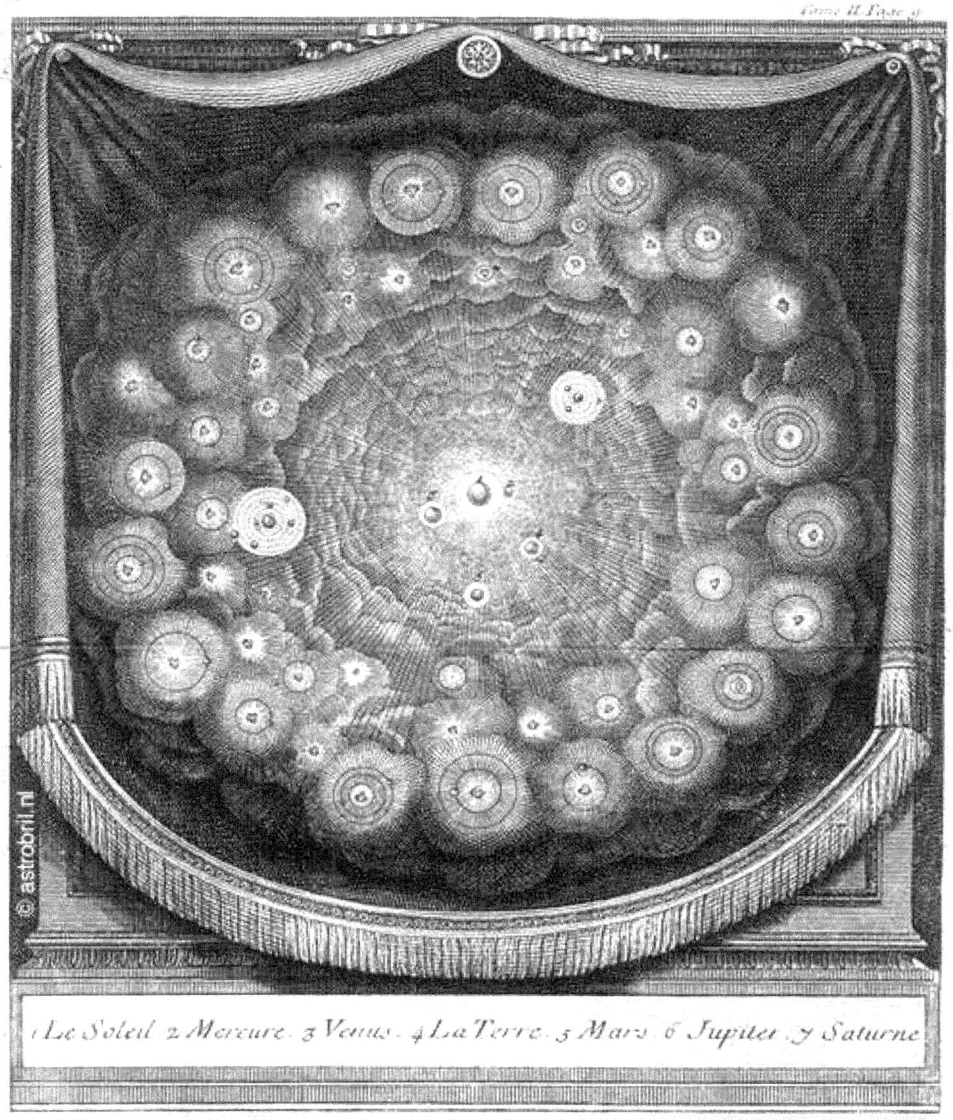

Figure 3.2. The frontispiece from an 1821 edition of Fontenelle's "Plurality of Worlds" published by Steven Dick. The illustration found in Fontenelle's first edition of "Entretiens dur la pluralite des mondes" in 1686, which clearly shows planets circling fixed stars, has been updated in this 1821 version to include the planet Uranus, discovered by William Herschel in 1781. Image obtained from SI libraries Dibner Collection.

universe. And the question still remains, are we alone? In that same 4th century BC when Aristotle was proclaiming our uniqueness (Chapter 2), the philosopher Epicurus was postulating the presence of infinite worlds. In 1658 Bernard de Fontenelle (1657–1757) said, "If God glorified Himself in making one world, the more worlds He made the greater must be His glory" (Figure 3.2). But perhaps only since the space age,

Figure 3.3. As part of the first ever "portraits" of the solar system taken by Voyager 1 in 1990 this image shows our Earth at a distance of 4 billion miles. Referred to by Carl Sagan (1934–1996) as "the pale blue dot" Earth appears to be in a beam of sunlight: an effect caused by photographing close to the Sun. (Reproduced courtesy of NASA/JPL.)

Figure 3.4. This Cassini image of Saturn backlit by the Sun was taken on 2013 July 19. It includes another image of our Earth (right of center below the inner bright rings) that truly brings home Carl Sagan's point: "We succeeded in taking that picture, and, if you look at it, you see a dot. That's here. That's home. That's us. Everyone you ever heard of, every creator and destroyer of civilizations, every king and peasant, every young couple in love, every hopeful child, every saint and sinner in the history of our species, lived there on a mote of dust, suspended in a sunbeam." (Sagan, 1999). Covering 405,000 miles across Saturn and its inner rings the image also captures Venus, Mars, and several of Saturn's more than 80 moons. (Reproduced courtesy of NASA/JPL-Caltech/SSI.)

when Carl Sagan referred to our world as a "pale blue dot" (Figures 3.3 and 3.4) in a vast universe, has the question taken on such a poignant appeal (Sagan 1973).

The search for other planets relates to some very basic but unquestionably opposing human needs: our desire not to be alone and our desire to be unique. By finding moons orbiting Jupiter, Galileo was the first to establish, beyond the shadow of a doubt, that our planet is NOT unique; worlds other than ours are also orbited by celestial bodies. But the question of whether we are alone remains.

3.2 Our Solar System

For hundreds of years our theories of planet formation were based on the only planetary system we were familiar with: our own. In our solar system, all planets orbit the Sun counterclockwise as seen from above Earth's North Pole, all have nearly circular orbits,

Figure 3.5. An artist's conception of the Beta Pictoris system, a site of active planetary formation: a newly formed star with its debris disk coalescing into planets. (Reproduced courtesy of NASA/FUSE/Lynette Cook.)

and all lie in approximately the same plane as the Earth. The theory that developed held that planet formation is essentially the end product of star formation. The interstellar gas cloud from which the Sun formed may have started off with only a slight rotation, but as its radius decreased through gravitational collapse (with essentially no change in mass) its rotation period had to increase. In order to conserve angular momentum a debris disk formed around the star. The material in the disk would all be rotating in the same plane, in the same direction, in a circular orbit around the Sun, as the planets in the solar system do today. The spinning of the gas and dust flattens out the disk in the equatorial plane of the star and eventually coalesces as planets (Figure 3.5).

Nebular gas by itself will not clump: but gas can condense onto dust particles and build all the way up to planets and stars. It takes a very, very long time. The process is similar to the way raindrops form: dust and soot in the air act as condensation nuclei on which water molecules adhere. As clumps grow larger their surface areas increase and so they can gather more material. The size of the planet then depends on its distance from the star: inner planets collect from a lesser area of debris, outer planets collect from a greater area. Why don't raindrops and planets continue to gather material and just get bigger and bigger? Raindrops get too heavy for air to support; and planets run out of material to accrete.

Planets could not form in the disk too close to the star, because the material would be too hot to coalesce, nor could they form too far from the star, because the available material would be too sparse. Even as ice solidifies from water at 273 Kelvin (0 degrees Celsius), different materials solidify at different temperatures. Another way to think about it is that different materials melt at different temperatures. Metals require the highest temperatures to melt so they form the closest to the Sun. In our solar system the temperature at which ice forms is reached at a distance 4 times that of the Earth from the Sun. The molecular clouds from which stars form are very, very cold:

Figure 3.6. Upper left (a): Mercury observed with the NASA/Messenger satellite. False color used to heighten contrast. The tan colored feature slightly left of center top is the Caloris basin created by an early impact and subsequently flooded with lava from volcanic activity; the light blue and white rays are material blasted out by recent impacts; the darker blue is low reflectance material. (Reproduced courtesy of NASA/JHU APL/ Carnegie Institution of Washington.) **Upper right (b):** In the early 1990s, radar imaging with the Venus orbiting Magellan spacecraft produced the first information on the planet's surface structure. In this computer generated picture of the radar data colors are based on images obtained by the Soviet Venera 13 and 14 landers. The bright area running across the middle is the largest highland region known as Aphrodite Terra. (Reproduced courtesy of JPL, NASA.) **Lower left (c):** Earth as photographed from Apollo 17 in 1972. Dubbed "Blue Marble" it is thought to be one of the most widely distributed photographs of any kind. (Reproduced courtesy of NASA; Photographic Mosaic & Digital Copyright: Rob Stevenso.) **Lower right (d):** Mosaic (composed of 102 Viking Orbiter images) of the Valles Marineris hemisphere of Mars projected into point perspective a distance of 2500 km above the surface of the planet. The dark area in the extreme north of this picture is Acidalia Planitia. The three Tharsis volcanoes (dark red spots), each about 25 km high, are visible to the west. (Reproduced courtesy of Jody Swann/Tammy Becker/Alfred McEwen, using the Planetary Image Cartography System developed at the U.S. Geological Survey in Flagstaff, Arizona.)

they have temperatures of about 10 degrees Kelvin (−263 degrees Celsius). Of course, as more matter accumulates, the cores of planets heat up from pressure. Also the heavier material sinks to the center (differentiation). Hence the planets nearer to the Sun are small, warm, and solid (Figure 3.6) whereas the planets farther from the Sun are large, cold, and gaseous (Figure 3.7). The planets are spherical for the same reason

Figure 3.7. Upper left (a): True color image of Jupiter constructed from four images obtained with the Cassini satellite in 2000. (Reproduced courtesy of NASA/JPL/University of Arizona.) **Upper right (b):** True color image of Saturn and its largest moon, Titan, obtained with Cassini. The rings are seen edge-on with their shadows on the planet surface. (Reproduced courtesy of NASA/JPL-Caltech/Space Science Institute/J. Major.) **Lower left (c):** Image of Neptune obtained in 1996 by Voyager 2. Neptune, like Uranus, is composed mostly of liquid water, methane, and ammonia and is surrounded by a thick gas atmosphere of mostly hydrogen and helium. (Reproduced courtesy of NASA, Voyager 2, Copyright Calvin J. Hamilton.) **Lower left (d):** An infrared image (in false color) of Uranus showing cloud bands, rings, and moons. Taken with the NICMOS camera on Hubble Space Telescope. (Reproduced courtesy of NASA/MSFC/HST.)

that stars are spherical: gravity. The minimum radius of an object that has acquired enough matter to become spherical is about 300km.

This theory of the formation of the solar system is supported by the fact that we find planets forming in the same places that stars form. The most nearby and best known star-forming region in our Galaxy, the Orion Nebula, contains approximately 700 protostars in various stages of evolution. It also contains approximately 180 identified protoplanetary systems, large disks of gas and dust surrounding newborn stars where planets are being formed. Figure 3.8 shows the location of several of these "proplyds" within the nebula and Figure 3.9 shows some observed characteristics (Ricci et al. 2008; O'Dell et al. 2009). The number of proplyds found

Figure 3.8. The great nebula in Orion, the nearest star-forming region, hosts hundreds of protoplanetary disks. Six of these (132-1832, 206-446, 180-331, 106-417, 231-838, 181-825) are highlighted in this image. The disks near to the bright star shine brightly. The disks that are farther from the radiation of the star show as dark silhouettes. Brighter disks are indicated by a glowing cusp in the excited material facing the bright star at random orientations within the nebula, so some appear edge on, and others face on. Interesting features include emerging jets of matter and shock waves. The shock waves are formed when the stellar wind from the nearby massive star collides with the gas in the nebula, sculpting boomerang shapes or arrows. (Reproduced courtesy of NASA, ESA/Hubble, M. Robberto (STScI/ESA), the HST Orion Treasury Project Team, & L. Ricci (ESO).)

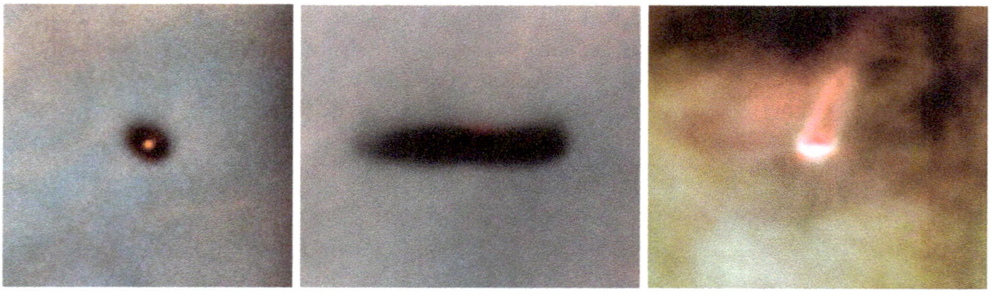

Figure 3.9. Close-up images of some protoplanetary disks (proplyds) in the Orion Nebula. **Left (a):** A proplyd showing a dim star (glowing in infrared) surrounded by a dusty disk visible against the brightness of the nebula. **Center (b):** A proplyd seen edge-on. **Right (c):** A proplyd showing a cometlike tail caused by a wind from the protostar blowing away the dust disk. (Reproduced courtesy of NASA, ESA, M. Robberto (STScI/ESA), the HST Orion Treasury Project Team, & L. Ricci (ESO).)

Figure 3.10. An artist's concept of the planetary system around the star Kepler-62 (discovered with the Kepler space telescope) compared to the inner planets around our Sun. Kepler-62 is a K2 dwarf, roughly two-thirds the size of our Sun and only 1/5th as bright. It has planets much larger than Earth on orbits closer to their Sun than Mercury is to ours. Green disks represent the habitable zones. The sizes are to scale. (Reproduced courtesy of NASA Ames/JPL-Caltech.)

in the Orion Nebula are a good indication that planet formation is fairly common in our universe.

The large disks of dust that show up in infrared observations of young stars and star-forming regions are the material out of which planets are formed. The material ejected when a star explodes and throws out its outer shell is the stuff of which stars, planets, and people are made. Although a star is a minute fraction of the mass of a galaxy, rates of star formation effect galaxy formation. Similarly, although planets are but a minute fraction of the mass of a star, perturbations due to planets can have a significant impact on the birth and death of stars: the intricate and complex patterns seen in planetary nebulae are thought to be caused by the interactions of the debris flung out from a collapsing star with the interstellar medium and planets in the system.

The revolution caused by the detection of planets outside our solar system (exoplanets) is that they do not look anything like us. For example, they can have very massive planets close to the star (Figure 3.10), their orbits can be highly elliptical (Figure 3.11), they can be massive and solid or low-mass and gaseous (Figure 3.28). In the past decade our theories of planet formation have had to be significantly revised. Theorists have been kept very busy trying to produce physical models of the unusual planetary characteristics for the new solar systems.

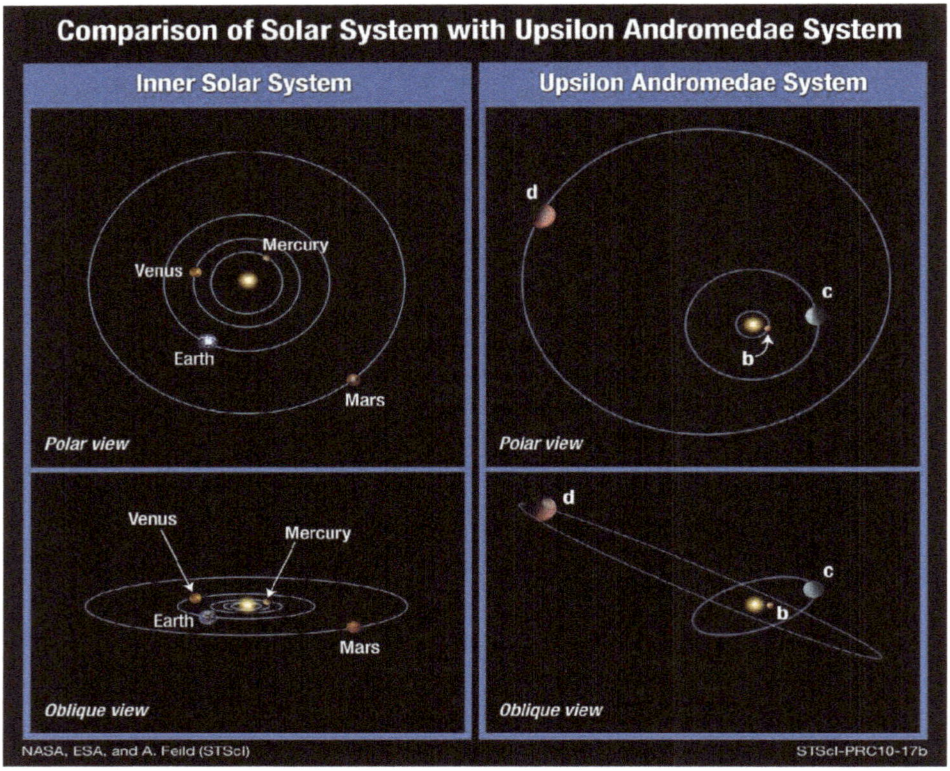

Figure 3.11. This artist's view of exoplanets around Upsilon Andromedae show significantly greater ellipticity than any planet in our solar system. (Reproduced courtesy of NASA, ESA and A. Feild (STScI).)

3.3 Exoplanets

Planets are small: the diameter of the Earth is less than one hundredth the diameter of the Sun and in visible light Earth is one billion times fainter than the Sun (Figure 3.12). The discovery of planetary systems around stars other than our Sun is a good example of synergy among studies in different wavelength bands and the use of innovative techniques. Methods used for detecting planets include radio timing of pulsars, imaging using adaptive optics in optical and infrared, radial velocities, or transit timing in optical and infrared. X-ray, ultraviolet, and polarimetry observations have also been used to determine the nature of exoplanets.

The first planet outside our solar system was detected in 1991 by Alexander Wolszczan of Pennsylvania State University and Dale Frail of NRAO (Wolszczan & Frail 1992). Wolszczan & Frail used the world's largest radio telescope at the time (Arecibo) to time the radio signals coming from a pulsar in the constellation Virgo, about 400 parsecs (1.2×10^{16} km) from Earth. Radio pulsars, especially those with a very rapid spin, represent the most precise natural clocks in the universe. The rotating beams of radio emission, rigidly attached to the star create a geometry that is analogous to that of a beam of light emerging from the lighthouse

Figure 3.12. Images of Earth and the other planets is superposed on this SOHO image of turbulent activity on the Sun. A large loop of superheated gas is erupting from the upper right. (Credit: SOHO-EIT Consortium, ESA, NASA.)

(see Chapter 4: New Actors on the Cosmic Stage). As a result, the observer records regular flashes or pulses of radio emission appearing periodically as dictated by the neutron star spin. These periodic pulses can, in fact, be favorably compared to the ticking of the best atomic clocks on Earth. If a rapidly spinning pulsar is orbited by a planet, the pulsar appears to wobble around the center of mass of the combined system. Because the pulsar does not stand still in space, its pulses arrive at the telescope with a tiny, variable delay that perturbs the ticking of our pulsar clock in a measurable way. Millisecond departures from the clock's regular behavior can be caused by Earth-sized planets and are easy to detect with an atomic comparison clock. In fact, even microsecond clock irregularities can be measured with the pulse timing technique, which means that it has the capability to detect large asteroids! Wolszczan & Frail determined the presence of two planets that are similar in mass to Earth and one that is about the mass of the Moon. These three initially discovered exoplanets have masses of 0.02, 4.3, and 3.9 Earth masses and their orbits are inclined at about 50° with respect to the plane of the sky. Compared to our solar system, the three planets (which include the smallest planet detected) would fit within the orbit of Mercury with the respective orbital periods of 25, 66, and 98 days (Figure 3.13; Wolszczan 1994).

There is some evidence that the pulsar may have an asteroid belt that appears to be located well beyond the orbit of Mars, just like it happens in our solar system. In addition, gravitational perturbations between the two larger planets have been detected and then used to measure true masses and orbital inclinations of these

Figure 3.13. An artist's comparison of sizes and distances of the planetary system found by Wolszczan & Frail around the pulsar B1257 + 12 in 1991. The orbits of all three planets would fit within the orbit of Mercury in our solar system, although two of the planets are significantly larger than Earth. (Reproduced courtesy of Pablo Carlos Budassi.)

objects, as noted above. The fact that the planets have almost coplanar orbits is convincing evidence that they have evolved from a protoplanetary disk in the process that was probably similar to the one that created planets around our Sun. The planets found are unlikely to support life because the tiny pulsar star they orbit bombards them with deadly radiation. Planets around pulsars are very rare and to date only about a half-dozen pulsars are confirmed to have planets around them (out of thousands of pulsars in our Galaxy).

One of the nearest examples of an exoplanetary system still forming is around the star Beta Pictoris, which is at a distance of 2.1×10^{14} km (22 light years). In the early 1980s, both ground-based and satellite observations showed evidence of a large disk around the star as well as an inner clear region roughly the size of our solar system. Images were made by blocking the light from the star itself to allow the fainter emission from the disk to show up. The warped appearance of the disk was indirect evidence for the presence of a disturbing body, most likely a planet.

Optical images and ultraviolet spectroscopy obtained with HST showed the warps of Beta Pictoris in great detail (Figure 3.14; Heap et al. 2000). The disk of Beta Pictoris is huge compared to the size of our solar system; greater than 1300 time

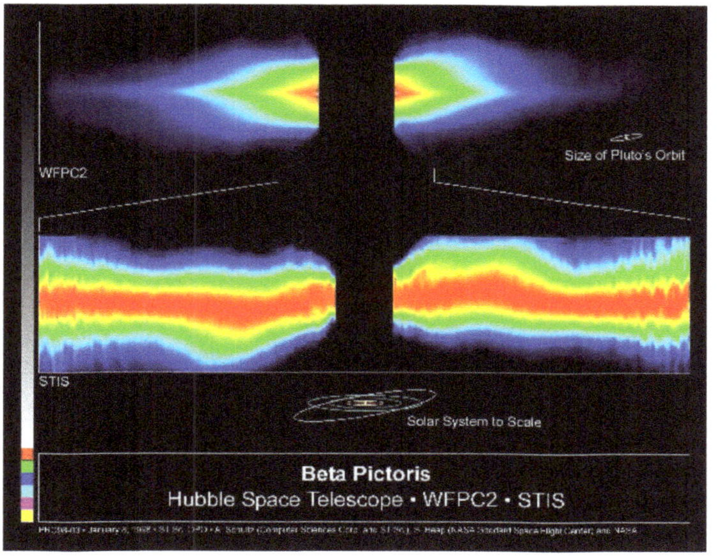

Figure 3.14. **Upper panel:** An optical image taken with the WFPC2 camera on Hubble/WFPC2 in 1995 shows the full extent of the 225 billion km disk around the star Beta Pictoris. A flaring at the top of the right side of the disk is caused by dust that has been pulled above the dense plane of the disk. **Lower panel:** A detailed view of the inner 2.3 billion km of the disk obtained with the Hubble/STIS in 1997 shows the warp in the disk clearly. These details support the presence of one or more planets orbiting the star. (Reproduced courtesy of NASA/ESA, A. Schultz (Computer Sciences Corp.), S. Heap (NASA/ESA Goddard Space FlightCenter).)

the distance from the Earth to the Sun. The morphology is asymmetric, presumably due to gravitational perturbation of a small companion star. The disk is composed of millimeter-sized grains that add up to a few of Earth's moons. The inner disk is warped, suggesting the presence of one or more planets roughly the size of Jupiter. However, gravitational perturbation by an unseen substellar-mass companion farther from the star than planets would be, or a tug from a bypassing star, might also cause this flaring. Using adaptive optics (a technique that can correct for atmospheric distortions) at infrared wavelengths, scientists at the European Southern Observatory were able to image the exoplanet itself (Figure 3.15; Lagrange et al. 2009).[1] In 2003 the exoplanet was seen on one side of the central star, its orbit then took it too close to the star to resolve, its detection in 2009 on the other side of the star confirmed its nature as a exoplanet.

The first direct images of exoplanets by observers using telescopes both on the ground and in space were announced simultaneously in 2008 (Marois et al. 2008; Kalas et al. 2008). Both were obtained using infrared wavelengths: this is because relative to their stars, planets are brighter in the infrared than in the optical. For example, the Earth is one billionth the brightness of the Sun in optical light, it is only one millionth the brightness of the Sun in infrared. The remarkable achievement from the ground,

[1] Named Beta Pictoris b after a convention in which the host star is a and planets orbiting are named b, c, d, and so on in order of discovery.

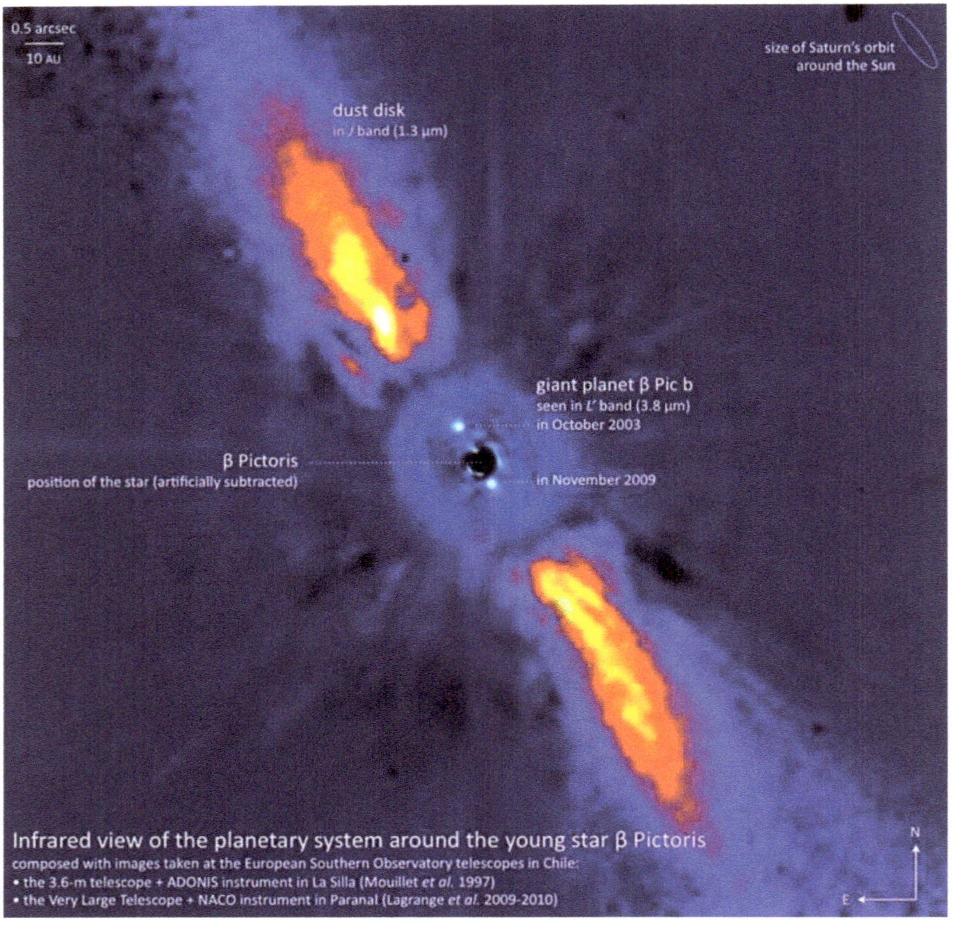

Figure 3.15. The exoplanet Beta Pictoris b was first imaged with the European Southern Observatory's Very Large Telescope in 2003 and confirmed in 2009. As shown by the sketch on the upper right the orbit of Beta Pictoris b around its star is similar in size to that of Saturn's orbit around our Sun. (Reproduced courtesy of A.-M. Lagrange, D. Ehrenreich (LAOG), et al., NASA/ESO.)

something no one had anticipated even a decade earlier, was possible by the use of adaptive optics (AO). On the ground, one of the main deterrents to seeing fine details is the distortion introduced by the Earth's atmosphere. Called astronomical "seeing" the best telescope sites on the ground at the best times can resolve to about 1 arcsec. An idea for compensating for the distortion was floated as early as 1953 (Babcock 1953), but did not become feasible until the advances in computer technology in the 1990s. The idea is to have a known steady source that is nearby the target to the extent that atmospheric turbulence should be similar between the steady source and the target. A deformable mirror is adjusted such that the correct image of the steady source is detected, using the adjusted mirror to observe the target provides an image "corrected" for the atmospheric distortion. The need for a steady source nearby relied on another technical advance: the

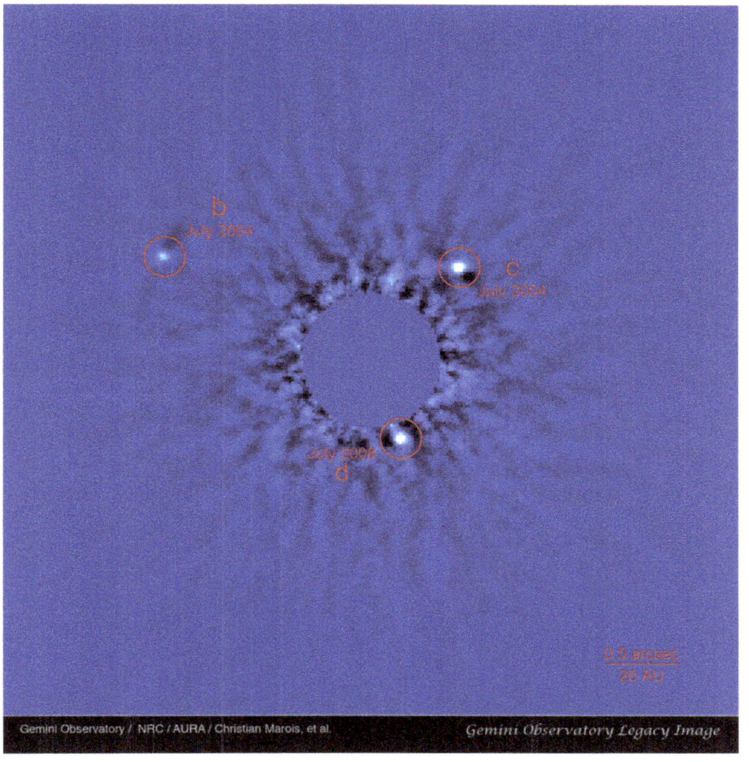

Figure 3.16. Exoplanets around HD 8799 observed with Gemini/Altair/NIRI telescopes using adaptive optics. The light from the central star was artificially obscured in order to allow the weaker signal from the planets to be detected. Multiple observations show the planets in Keplerian orbits around the HD 8799. (Reproduced courtesy of Gemini Observatory/NRC/AURA/Christian Marois et al.)

ability to generate artificial comparison stars using lasers. This technique works by using lasers to excite atoms in the atmosphere producing a guide "star." The higher up in the atmosphere the guide star is produced the more of the atmosphere can be compensated. Earth's mesosphere (at 90 km) and thermosphere contain sodium. By shining a laser of the right wavelength (589 nm), these sodium atoms can be excited and appear to glow. AO can get "seeing" on the ground down to 0.1 arcsec. Not as good as HST which gets 0.05 arcsec but with the collecting power for say a 10 m telescope resolution is up a factor of 20.

Christian Marois of the Herzberg Institute (Canada) and collaborators (Marois et al. 2008) observed the star HR 8799 for several years with the 8.2 m Gemini North telescope, following two planets. They confirmed their detection using observations with the 10.0 m Keck II telescope. Their Keck II observations showed the presence of a third planet much closer to HR 8799. Figure 3.16 shows an image obtained with Gemini North in September of 2008 and published in the 2008 November 13 issue of Science Express. HR 8799 has a mass about 1.5 times that of our own Sun, and lies about 130 light years from the Sun, a distance similar to many stars easily visible in

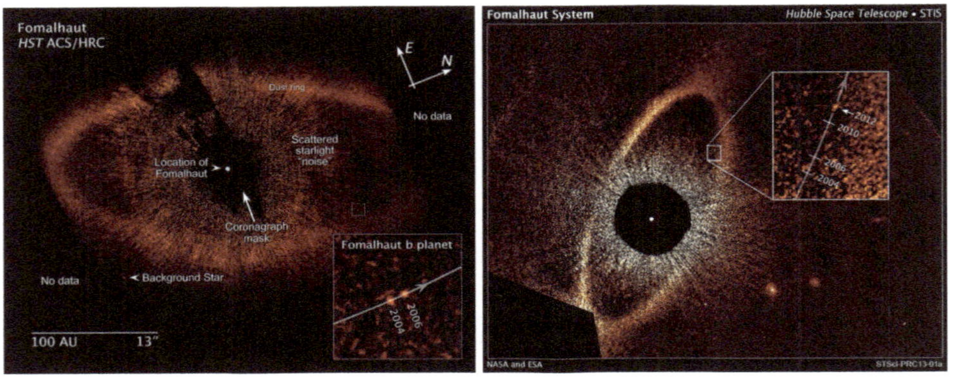

Figure 3.17. Exoplanet around Fomalhaut. The black circle at the center of the image blocks out the light from the bright star, allowing reflected light from the belt and planet to be photographed. **Left (a):** Hubble/ACS images of Fomalhaut showing its dusty debris ring and planet as detected in 2004 and 2006. This represents the first visible-light image of a planet circling another star. (Reproduced courtesy of NASA, ESA, P. Kalas, J. Graham, E. Chiang, E. Kite (University of California, Berkeley), M. Clampin (NASA/Goddard), M. Fitzgerald (Lawrence Livermore NL), K. Stapelfeldt, J. Krist (NASA/JPL).) **Right (b):** Hubble/STIS images in 2012 show the planet continuing on the orbit determined in 2008. (Reproduced courtesy of NASA, ESA, and P. Kalas (University of California, Berkeley and SETI Institute).)

the night sky. Although the HR 8799 planetary system has significant differences with our solar system, it is a clear demonstration that complex planetary systems exist, systems that could conceivably contain Earth-like planets. The HST/ACS discovery announced by Paul Kalas and colleagues (Kalas et al. 2008) was based on more than two years of observations of a star called Fomalhaut (Figure 3.17(a)). Observations with HST/STIS in 2012 confirmed the detection and verified the orbit of the planet (Figure 3.17(b)). Based on these observations and additional observations with ALMA (Boley et al. 2012) astronomers calculated that the planet, called Fomalhaut b, is in a 2000-year-long, highly elliptical orbit (Figure 3.18). The planet will appear to cross a vast belt of debris around the star roughly 20 years from now (2022). If the planet's orbit lies in the same plane with the belt, icy and rocky debris in the belt could crash into the planet's atmosphere.

The single most fruitful method for finding extra-solar planets from the ground is through measuring their radial or Doppler velocities. This method uses the gravitational influence of the planets on their host star and therefore is best at detecting very massive objects close to the parent star—so-called "hot Jupiters,"—because they have the greatest gravitational effect on the host star, and so cause the largest changes in its radial velocity. As the star moves away from us light waves are shifted toward the red end of the spectrum: as they move toward us they are shifted to the blue (Figure 3.19). The shifts are very small and a pioneer design exploited by Geoff Marcy and collaborators at San Francisco State University involved putting iodine vapor in a glass cell in front of the slit of a spectrograph (Marcy & Butler 1992). As star light passes through the iodine, certain signature wavelengths are absorbed producing dark lines along the spectrum. These lines serve as reference for the

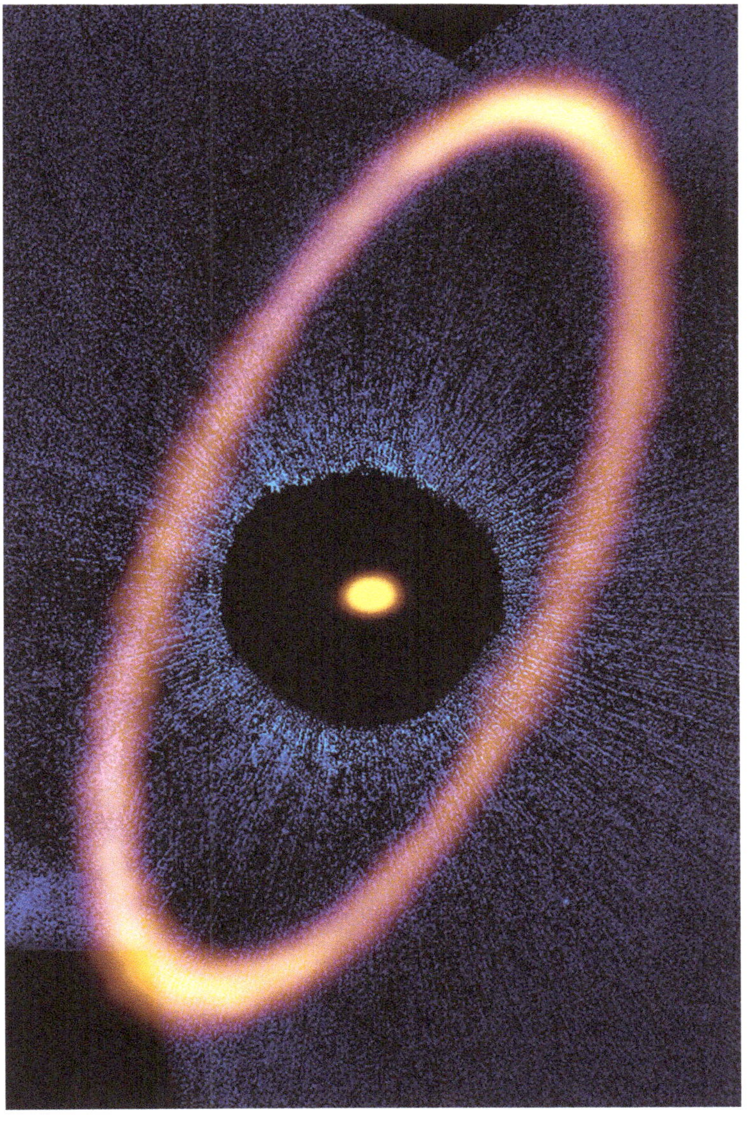

Figure 3.18. The Fomalhaut star system as imaged in 2012. ALMA data in orange show an eccentric debris disk. The yellow dot at the center is the unresolved star. The dark region was a coronagraphic mask applied to filter out light from the central star with the Hubble Space Telescope optical data depicted in blue. (Reproduced courtesy of ESO, ALMA (ESO/NAOJ/NRAO)/L. Matrà/M. A. MacGregor.)

Doppler shift, which is then translated to a radial velocity curve from which the existance of any planetary companions is assessed. With this innovation the Doppler technique is so sensitive that shifts in wavelengths as small as 1 part in 100 million can be discerned. The radial velocity method applies to main sequence stars of spectral types mid-F through M. Stars hotter and more massive than mid-F rotate

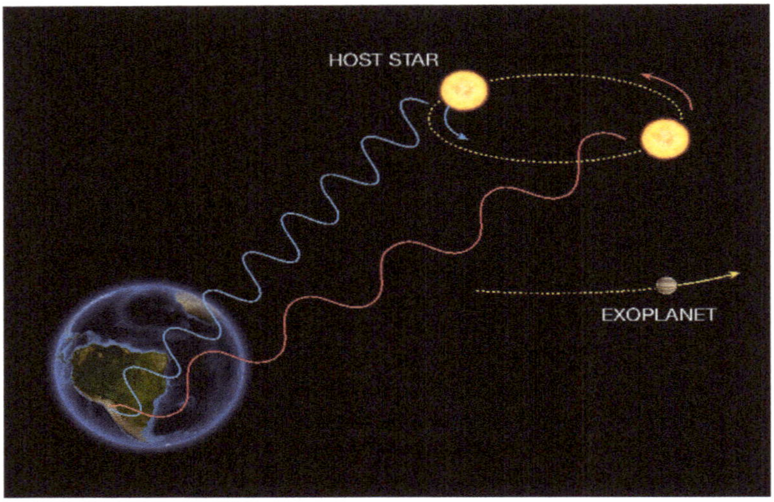

Figure 3.19. An animation of the radial velocity or Doppler shift method for detecting exoplanets. All planets have a gravitational effect on their star. As the planet moves around the star the star shifts position and wobbles back and forth. Spectral lines emitted by the star move to the right or red part of the spectrum as the star goes away from the observer and to the left or blue part of the spectrum as it approaches. (Reproduced courtesy of NASA/ESA/ESO.) Animation available at http://doi.org/10.1088/2514-3433/ac81b3.

faster, pulsate, are generally more active and have less spectral structure, thus making their Doppler shift more difficult to be measured. A further problem is that the gas envelope around certain types of stars can expand and contract, and some stars are variable. This method is unsuitable for finding planets around variable stars as changes in the stellar emission spectrum caused by the intrinsic variability of the star can swamp the small effect caused by a planet. Observations of many separate spectral lines and many orbital periods increases the chance of observing smaller and more distant planets, but Earth-size planets remain undetectable using this method with current instruments.

A major problem with the Doppler method is that it can only measure movement along the line-of-sight, and so depends on a measurement (or estimate) of the inclination of the planet's orbit to determine the planet's mass. If the orbital plane of the planet happens to line up with the line-of-sight of the observer, then the measured variation in the star's radial velocity is the true value. However, if the orbital plane is tilted away from the line-of-sight, then the true effect of the planet on the motion of the star will be greater than the measured variation in the star's radial velocity, which is only the component along the line-of-sight. As a result, the planet's true mass will be higher than expected. To correct for this effect, and so determine the true mass of an exoplanet, radial-velocity measurements can be combined with astrometric observations, which track the movement of the star across the plane of the sky, perpendicular to the line-of-sight. Astrometric measurements allow researchers to check whether objects that appear to be high mass planets are more likely to be brown dwarfs.

The first planet around a sunlike star was discovered in 1995 by Michel Mayor and Didier Queloz of the Geneva Observatory in Switzerland using the Doppler method (Mayor & Queloz 1995). The planet they found, 51 Pegasi b, was about half the mass of Jupiter (147 times the mass of the Earth) and at a distance 200 times closer to its star in an orbit less than one-sixth the distance from our Sun to Mercury (Figure 3.20).

Within only a few years of first detecting extra-solar planets we were able to detect their atmospheric temperatures and composition.

Figure 3.21(a) shows the temperature distribution on the surface of HD 189733b as measured by Sptizer's infrared array camera (Knutson et al. 2007). Called a hot-Jupiter planet, because it is of a size comparable to Jupiter but is located extremely close to its star. HD 189733b completes one orbit around its star in just 2.2 days compared to Mercury's 88 day orbit around our Sun. Spitzer observed the object for half of its orbit. This likely means that one side of the planet always faces its star. Figure 3.21(a) shows HD 189733b has a warm spot on the side that always points toward the star, however it is somewhat offset from center. The suggested explanation is that there are very high winds (nearly a 1000 km per hour) that push the hot material over. There is less than a factor two temperature difference between the sunlit and Sun dark side as one would expect from strong winds that help to distribute heat rapidly.

The atmosphere of another hot-Jupiter planet, HD209458b, is actually blown away by the strong stellar wind particles from its extremely hot star (Ballester et al. 2007; Figure 3.21(b)). HD 209458b is directly in our line of sight to its star. As it passes in front of the star, the dimming of its optical light helps to determine the structure of its atmosphere. Spectroscopic studies with the STIS on Hubble show absorption features in the different layers of the atmosphere. These indicate the presence of sodium (Charbonneau et al. 2002), oxygen and carbon (Vidal-Madjar et al. 2004), and hydrogen (Vidal-Madjar et al. 2003).

Spitzer observations of the exoplanet WASP-12b at four different wavelengths combined with ground-based observations at lower infrared wavelengths show molecules of carbon monoxide and methane in its atmosphere

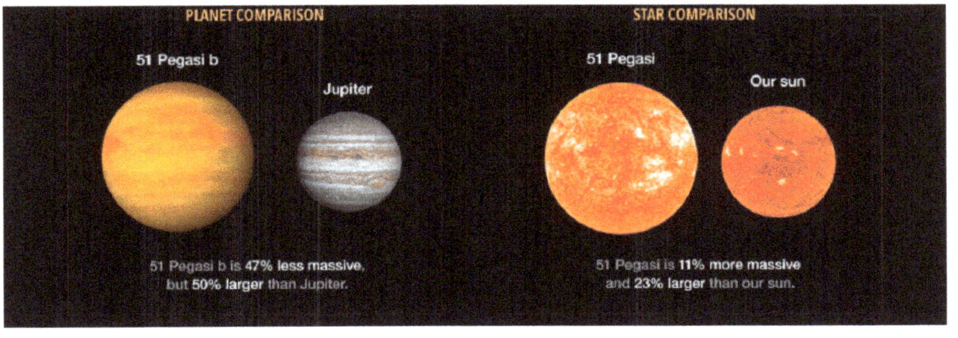

Figure 3.20. An artist's comparison of 51 Pegasi and its planet with our Sun and Jupiter. **Left (a):** Exoplanet 51 Pegasi b is somewhat larger in size but smaller in mass to Jupiter. **Right (b):** 51 Pegasi is both larger in size and mass compared to our Sun. (Reproduced courtesy of NASA/JPL-Caltech.)

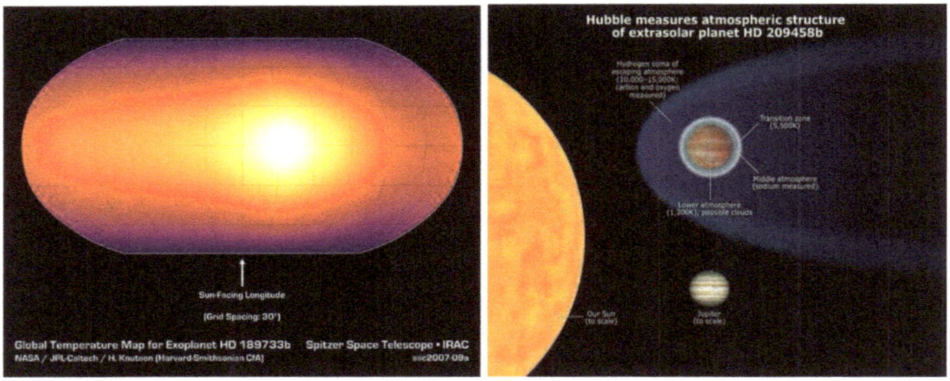

Figure 3.21. Left: A map of the surface of a planet beyond our solar system. The map, which shows temperature variations across the top of a gas giant called HD 189733b, is made from infrared data taken by NASA's Spitzer Space Telescope. HD 189733b is thought to always face one side toward its star. Hotter temperatures are represented in brighter colors. The range is roughly 650 degrees Celsius on the cool side and 930 degrees Celsius on the hot side. (Reproduced courtesy of NASA/Spitzer.) **Right:** The extrosolar planet HD 209458b orbits close to its star and is so hot that material from its atmosphere is blown away by the strong stellar irradiation. (Reproduced courtesy of NASA, ESA, and A. Feild (STScI).)

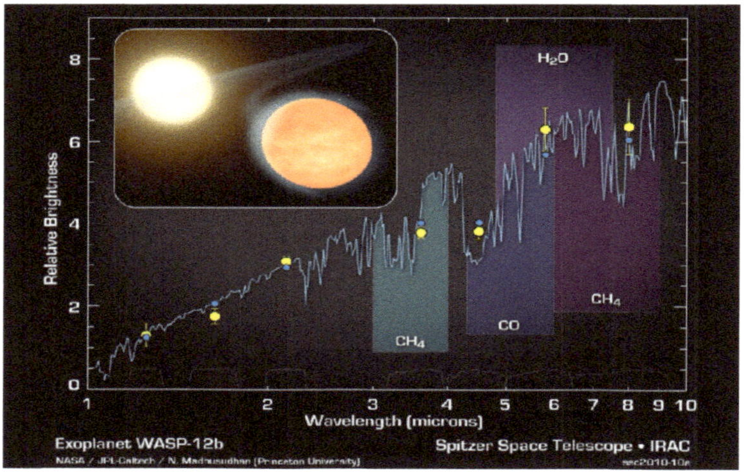

Figure 3.22. Spitzer Space Telescope observations of molecules in the atmosphere of the exoplanet WASP-12b, a super-hot gas giant that orbits very close to its star. The yellow dots show the data, along with the observational uncertainties. The blue curve is a model of the planet's light, or spectrum, showing the fingerprints of chemicals in the atmosphere. The blue dots represent the blue model curve averaged to cover the same wavelengths as the data, as shown by the gray lines at the bottom of the plot. (Reproduced courtesy of NASA/JPL-Caltech/CFHT/MIT/N. Madhusudhan [Princeton University]/UCF.)

(Figure 3.22; Madhusudhan et al. 2011). So although first carbon-rich exoplanet discovered, a positive sign for life, the atmosphere is very lethal. The atmospheric signature of the planet was obtained by observing the star and planet together and the star by itself while the planet was behind it. WASP-19 is another star that is orbited by a "hot Jupiter." Hubble STIS observations of the ultraviolet spectra of WASP-19b shows both methane and hydrogen cyanide in its atmosphere with little to no water (Huitson et al. 2013; Wong et al. 2016; Figure 3.23).

In 2009, NASA launched the Kepler satellite which was specifically designed to detect planets by measuring the reduction in light from a star caused by a planet passing in front of its star in our line of sight (the transit method, Figure 3.24). With the transit method one is able to measure both the mass and the size of the transiting planet. Kepler was designed to continuously monitor over 150,000 normal stars in a specific region of our Galaxy (Figure 3.25). The selected area is roughly at the same

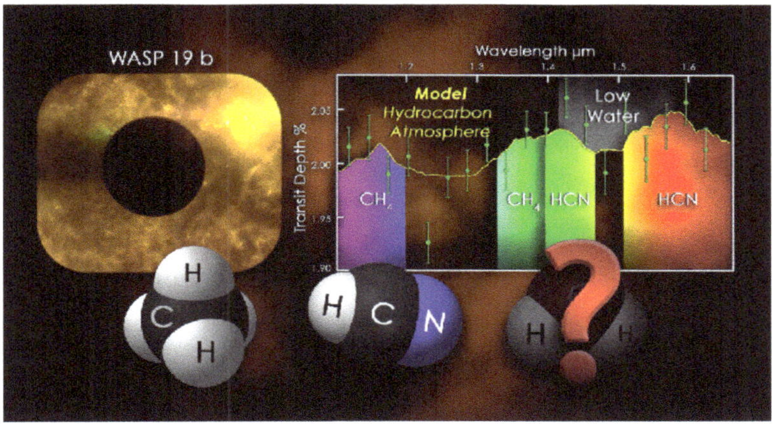

Figure 3.23. Hubble Space Telescope STIS spectra of exoplanet WASPP-19b shows high abundances of hydrocarbons like methane and hydrogen cyanide in the planet's thick atmosphere, and very little trace of water. (Reproduced courtesy of NASA/Hubble.)

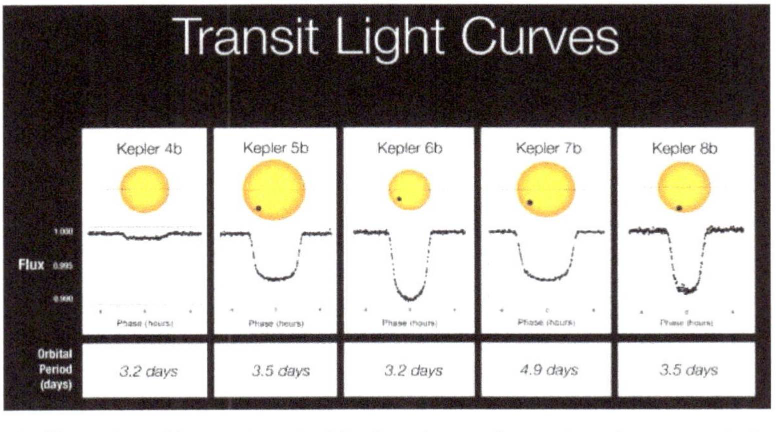

Figure 3.24. An illustration of the transit method for detecting exoplanets. As a planet passes in front of its star the light from the star is dimmed. The width of the dip is related to the size of the star and the amount of dimming is related to the size of the planet. (Reproduced courtesy of NASA/Kepler.)

Figure 3.25. Right (a): An artist's depiction of the Kepler search space. (Reproduced courtesy of NASA Kepler Mission, Jon Lonberg.) **Left (b):** The actual field of view against an image of our Galaxy. (Reproduced courtesy of NASA Kepler Mission, Carter Robertsr.)

distance from the center of our Galaxy as the Sun in order to take into consideration any effect that location in the Galaxy could have on the production of Earth-like planets.

3.4 Finding Earth-like Planets

The Earth is unique within our solar system as the only planet on which life is known to exist. We attribute this to the fact that Earth contains liquid water on its surface, and water is a good medium within which to grow the complex molecules that lead to the development of life. To have liquid water on its surface the temperature of a planet has to be between 0 and 100 degrees centigrade. Below this temperature range water freezes, above this range it boils away. The region of a solar system where planets can have liquid water on its surface is referred to as the "Habitable Zone."

The first exoplanet found in the habitable zone of its star was Gliese 581 c (Figure 3.26; Zollinger & Armstrong 2009; von Paris et al. 2010; von Bloh et al. 2007). The planets around Gliese 581 were detected with the High Accuracy Radial velocity Planet Searcher (HARPS) installed at the European Space Agency's 3.6 m telescope at La Silla Observatory in Chile (Vogt et al. 2010). At a distance of 1.9×10^{14} km (20 light years), Gliese 581 is the 89th closest star to our Sun. It is a red dwarf star about 1/3rd the size of the Sun. Since Gliese 581 is smaller and has a lower surface temperature than our Sun, its habitable zone is much closer to the star. Gliese 581c, first detected in 2007 is about five times the mass of Earth, a so-called "super-earth." In 2011, Gliese 581g at a mass twice that of Earth was located in the very center of the habitable zone (von Bloh et al. 2011).

The first Earth-size planet found within the habitable zone of its star was detected with the Kepler satellite in the first two years of its mission with their existence confirmed in February 2014 (Figure 3.27).

Figure 3.26. Gliese 581 is a red dwarf star one-third the size of the Sun. A comparison of the relative sizes of our solar system and the Gliese 581 system. Gliese 581c, discovered in 2007, was the first exoplanet discovered in a possibly habitable zone; in 2010 Gliese 581g and 581d were discovered to exist solidly in the habitable zone, with g somewhat smaller and d roughly twice the size of Earth. (Reproduced courtesy of NASA/Ames Research Center/Natalie Batalha/Wendy Stenzel.)

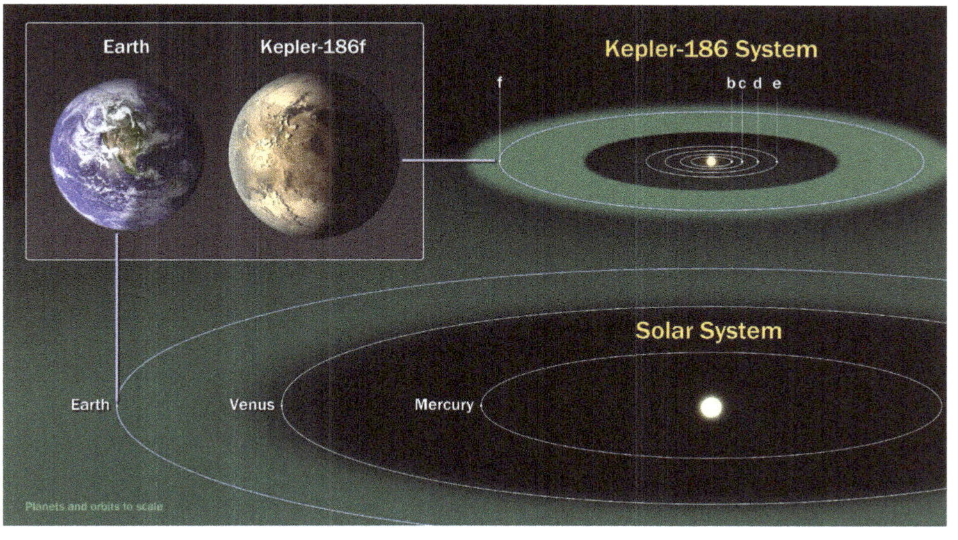

Figure 3.27. The Kepler 186 system is the first Earth-like planet found by the Kepler satellite in the habitable zone of its star. (Reproduced courtesy of NASA/ESO.)

In 2020 both the first extragalactic planet (M51-ULS-1b) and the first planet not associated with a star (earth-mass rogue planet) were reported. The first by detecting an eclipsing X-ray source in the Whirlpool Galaxy (M51) with the Chandra X-ray satellite (Di Stefano et al. 2021) and the latter by using microlensing techniques

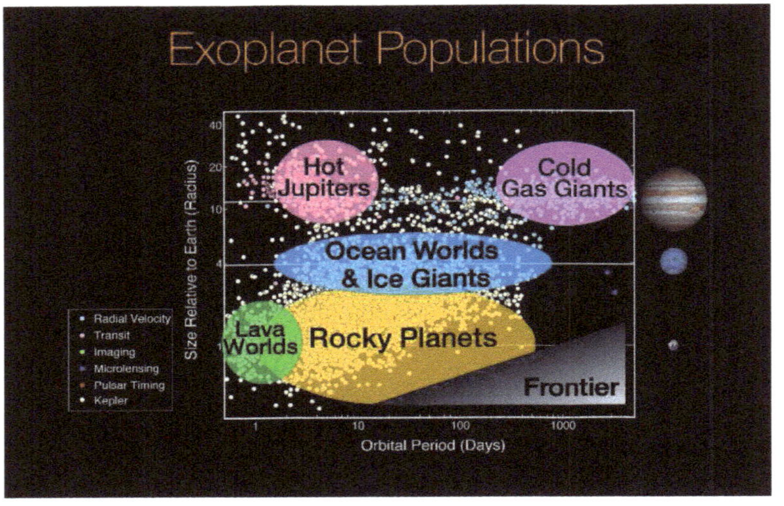

Figure 3.28. As of January 2020, 4374 exoplanet candidates has been detected by the Kepler and TESS missions. Less than 1 percent of these are in the habitable zone of their stars. (Reproduced courtesy of NASA Kepler and TESS Missions.)

(see Chapter 7, Figure 7.9) on the first Earth-sized exoplanet in a habitable zone detected by the Transiting Exoplanet Survey Satellite (TESS; TOI-700d; Gilbert et al. 2020; Rodriguez et al. 2020; Suissa et al. 2020).[2]

By its retirement in late 2018 of the over 5000 exoplanet candidates detected by Kepler, 2662 had been confirmed. The distribution of sizes and detection methods depicted in Figure 3.28 demonstrates Kepler's superior ability to detect Earth-sized exoplanets.

By November 2018, NASA had launched TESS. TESS also used the transit method but was designed to cover an area 400 times larger than that surveyed by Kepler and reaching the stars nearest to us. As of November 2021, TESS had found 4700 candidate exoplanets of which 172 have been confirmed. These are the likeliest objects for detailed investigations and will provide prime targets for the James Webb Space Telescope.

References

Babcock, H.W. 1953, PASP, 65, 386
Ballester, G. E., Sing, D. K., & Herbert, F. 2007, Natur, 445, 511
Boley, A. C., Payne, M. J., Corder, S., et al. 2012, ApJ, 750, L21
Charbonneau, D., Brown, T. M., Noyes, R. W., & Gilliland, R. L. 2002, ApJ, 568, 377
Di Stefano, R., Berndtsson, J., Urquhart, R., et al. 2021, NatAs, 5, 1297
Gilbert, E. A., Barclay, T., Schlieder, J. E., et al. 2020, AJ, 160, 116
Heap, S. R., Lindler, D. J., Lanz, T. M., et al. 2000, ApJ, 539, 435
Huitson, C. M., Sing, D. K., Pont, F., et al. 2013, MNRAS, 434, 3252

[2] TOI stands for TESS Object of Interest.

Kalas, P., Graham, J. R., Chiang, E., et al. 2008, Sci, 322, 1345
Knutson, H. A., Charbonneau, D., Allen, L. E., et al. 2007, Natur, 447, 183
Lagrange, A.-M., Gratadour, D., Chauvin, G., et al. 2009, A&A, 493, L21
Madhusudhan, N., Harrington, J., Stevenson, K. B., et al. 2011, Natur, 469, 7328
Marcy, G. W., & Butler, R. P. 1992, PASP, 104, 270
Marois, C., Macintosh, B., Barman, T., et al. 2008, Sci, 322, 1348
Mayor, M., & Queloz, D. 1995, Natur, 378, 355
O'Dell, C. R., Henney, W. J., Abel, N. P., Ferland, G. J., & Arthur, S. J. 2009, AJ, 137, 367
Ricci, L., Robberto, M., & Soderblom, D. R. 2008, AJ, 136, 2136
Rodriguez, J. E., Vanderburg, A., Zieba, S., et al. 2020, AJ, 160, 117
Sagan, C. 1973, The Cosmic Connection An Extraterrestrial Perspective (Garden City, NY: Anchor Press, Doubleday)
Sagan, C. 1999, Pale Blue Dot (New York: Ballantine Books)
Suissa, G., Wolf, E. T., Kopparapu, R. K., et al. 2020, AJ, 160, 118
Vidal-Madjar, A., Lecavelier des Etangs, A., Désert, J.-M., et al. 2003, Natur, 422, 143
Vidal-Madjar, A., Désert, J.-M., Lecavelier des Etangs, A., et al. 2004, ApJ, 604, L69
Vogt, S. S., Butler, R. P., Rivera, E. J., et al. 2010, ApJ, 723, 954
von Bloh, W., Bounama, C., Cuntz, M., & Franck, S. 2007, A&A, 476, 1365
von Bloh, W., Cuntz, M., Franck, S., & Bounama, C. 2011, A&A, 528, 133
von Paris, P., Gebauer, S., Godolt, M., et al. 2010, A&A, 522, 23
Wolszczan, A., & Frail, D. A. 1992, Natur, 355, 145
Wolszczan, A. 1994, Sci, 264, 538
Wong, I., Knutson, H. A., Kataria, T., et al. 2016, ApJ, 823, 122
Zollinger, R., & Armstrong, J. C. 2009, A&A, 497, 583

Chapter 4

New Characters on the Cosmic Stage

Artist's impression of gamma-ray explosion reaching Earth. Courtesy of NASA.

4.1 Introduction

The opening of new wavelength bands or channels of information lead to discoveries of entirely new classes of objects as well as providing new views of familiar objects. Examples of new classes include radio pulsars, anomalous X-ray pulsars, X-ray binaries, gamma-ray bursts (GRBs), and soft gamma-ray repeaters, each of which was first discovered in the wavelength for which it is named. In addition to being

doi:10.1088/2514-3433/ac81b3ch4

totally unexpected before they were discovered, these "new characters" all required observations at wavelengths other than the wavelength of their discovery as well as other channels of information, before their true nature could be determined. Radio pulsars, which provided us with indirect evidence for the existence of gravity waves required by Einstein's General Theory of Relativity, and GRBs, which could possibly be the most energetic phenomena in the universe, were both first detected in 1967. An explanation for radio pulsars was developed within months of their discovery, whereas we are only now, with the detection of associated gravitational waves and afterglows, beginning to understand the nature of GRBs in terms of more characters: collapsers, hypernovae, and kilonovae. Anomalous X-ray pulsars and soft gamma-ray repeaters both turn out to be highly magnetized neutron stars now commonly called "magnetars."

4.2 Radio Pulsars

Radio pulsars were discovered by Jocelyn Bell and Anthony Hewish of Cambridge University (Hewish et al. 1968). Radio astronomy had been around for almost four decades and Bell and Hewish were searching for quasars.[1] In 1967 Hewish realized that a new technique just discovered would be useful for searching for radio-selected quasars. Because of the diffraction of radio waves as they pass through the turbulent solar wind, the intensity of the radio emission undergoes an apparent fluctuation or "scintillation." Compact radio objects, e.g., quasars, showed more scintillation than extended sources.[2]

As described in Jocelyn Bell's delightful article "Petit Four" (Bell Burnell 1977), Hewish, Bell, and a few others constructed by hand a radio telescope that covered 4.5 acres and consisted of a thousand wooden posts connected by 120 miles of wire and cable (Figure 4.1). In addition to the expected scintillation from the quasars, Bell found what she called "scruff," little blips of radio emission that kept accurate sidereal[3] time.

The "scruff," which turned out to be radio pulsar emission, is very distinctive: a uniform series of pulses, precisely spaced with periods between a few milliseconds and several seconds. After a (very brief) consideration of little green men (LGM), Bell and Hewish ascribed the incredible regularity of these radio pulses to the rapid rotation of neutron stars.

[1] The word 'quasar' developed from a term 'quasi-stellar radio source,' originally meant to apply to certain strong, compact, radio sources which looked like stars at optical wavelengths but whose extremely large redshifts implied distances well beyond our Galaxy (see Chapter 5: From Energy to Matter). In 1964, in an article written for Physics Today, Hong-Yee Chiu (Chiu 1964), shortened the term to "quasar."

[2] This is analogous to the way that stars appear to "twinkle" more than planets. The reason is that stars are so far away that they appear as points of light on the sky, while planets have a finite size. The disk of a planet subtends a larger angle of sky than the size of the turbulence and so averages out the effects of the atmosphere.

[3] Observation shows that any specific star will cross directly overhead (on the meridian) about four minutes earlier every day. In other words, the day according to the stars (the sidereal day) is about four minutes shorter than the day according to the Sun (the solar day). The reason for this is that if we point directly toward the Sun at exactly twelve noon, after a full rotation of the Earth, we are pointing in the same direction in space, but since the Earth has also proceeded along its orbit, we are no longer facing the Sun. It takes another 4 minutes of rotation for us to again directly face the Sun.

Figure 4.1. The radio telescope at Jodrell bank that was constructed by Jocelyn Bell Burnell (pictured) in 1967 with fellow graduate students and advisor, Antony Hewish. Over a 1000 wooden posts covering 4.5 acres and connected by 120 miles of wire and cable. (Reproduced courtesy of Jocelyn Bell Burnell.)

The existence of neutron stars had been postulated by Walter Baade and Fritz Zwicky in a paper identifying explosions of supernovae with the collapse of massive stars (Baade & Zwicky 1934; see Chapter 2: From Dust...The Birth and Death of Stars). When a massive star explodes as a supernova, it leaves behind a core that is only about 20 km across but which contains as much matter as our Sun; the creation of the core triggers the explosive ejection of the outer parts of the star. The core is so dense that neutrons are the only form of matter that can exist in it. Hence the name "neutron star": the basic idea is that when the central part of the star fuses its way to iron, it can't go any farther because iron has the highest binding energy per nucleon of any element, so neither fusion nor fission of iron can release energy. Thus, the iron core just accumulates until it gets to about 1.4 solar masses (called the Chandrasekhar mass limit[4]; Chandrasekhar 1931) at which point the electron degeneracy pressure that had been supporting it against gravity is overpowered and the star collapses inward. At the very high pressures involved in this collapse, it is energetically favorable to combine protons and electrons to form neutrons plus neutrinos. The neutrinos escape after some scattering and the neutrons settle into a roughly 20 km in diameter sphere with neutron degeneracy balancing the inward push of gravity. Radio pulsars are quite common in our Galaxy and their

[4] Below this mass limit one can form a white dwarf, which avoids gravitational collapse by electron degeneracy.

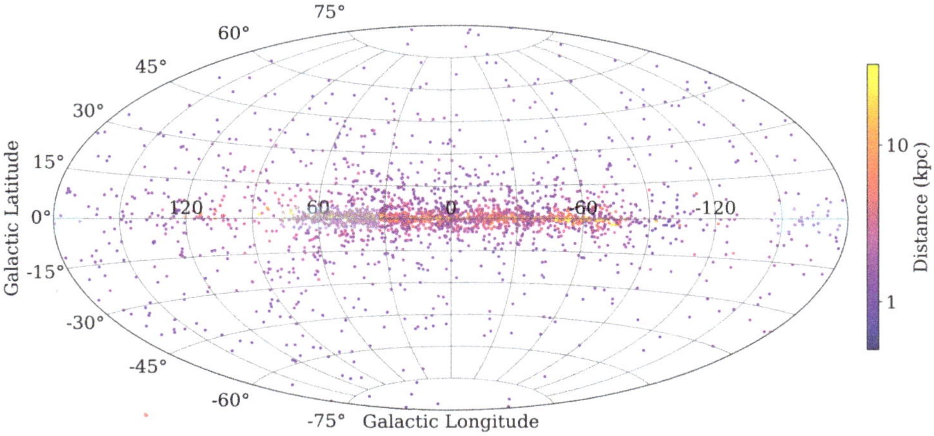

Figure 4.2. Sky map in Galactic coordinates. The dots represent known pulsars (circa 2019), color-coded by distance. The majority of the pulsars lie along the Galactic plane. (Reproduced courtesy of Emilie Parent.)

distribution is concentrated in the plane of our Galaxy, consistent with their identification with the deaths of massive stars (Figure 4.2).

A neutron star is so dense that, on Earth, one teaspoonful of neutron star matter would weigh a billion tons! In 1973, Paul Joss of MIT referred to them as "elementary particles with the mass of the Sun."[5] Because of their small size and high density, neutron stars possess surface gravitational fields about 10 billion times and surface magnetic fields about 1000 billion times that of a star like our Sun. This is because the strength of a surface magnetic field increases *inversely* with the surface area. Rapid rotation combined with such high magnetic fields makes pulsars powerful electric generators, capable of accelerating charged particles to energies of a thousand million volts. These charged particles, in a way not yet fully understood, are responsible for the beam of radiation in radio, optical, X-ray, and gamma-ray emission that comes out from each magnetic pole.

If the magnetic and rotational axes of a pulsar are misaligned, the radio beam sweeps around as the pulsar rotates, just as the spotlight in a lighthouse does (Figure 4.3). Like a ship in the ocean that sees only regular flashes of light from a lighthouse, we see pulsars turn on and off with strict timing as their beams sweep over the Earth (if the Earth happens to be in the path of the beam). By measuring how rapidly the beam flashes at Earth, we can determine a pulsar's rate of spin. The energy emitted by a radio pulsar comes from the rotation of the neutron star, which slows down as it loses energy. This slowing down is detected as a lengthening of the pulse period. Typically, a pulsar rotation rate slows down by several parts in a million each year.

The Crab pulsar is the first neutron star connected with a known supernova event (an explosion that was observed and recorded in 1054 AD by Japanese and Chinese

[5] Overheard by author at an MIT seminar in 1973.

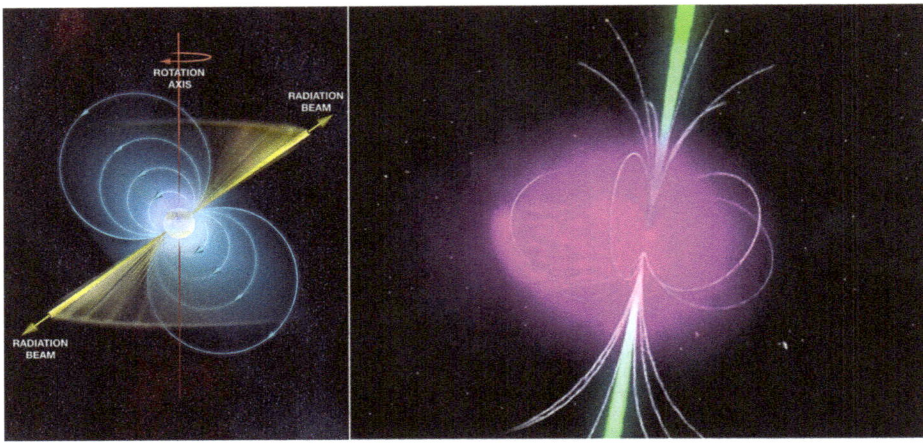

Figure 4.3. **Left (a):** Illustration of a pulsar showing the rotation axis (red) with direction of emission along the magnetic axis (yellow). Magnetic field lines are sketched in blue. (Reproduced courtesy of NRAO/AUI/NSF/Bill Saxton.) **Right (b):** Animation of a pulsar with the relatively narrow radio emission (green) moving in and out of our line of sight. Magnetic field lines are depicted in pink. The gamma-ray emission beam is represented in magenta; it is much broader and can be detected even when the radio beam is unfavorably oriented toward Earth. (Reproduced courtesy of NASA/Fermi/Cruz deWilde.) Animation available at http://doi.org/10.1088/2514-3433/ac81b3.

astronomers; see Chapter 2: From Dust...The Birth and Death of Stars, for further details on the Crab Nebula). First positively identified as a strong X-ray source in 1964 (Bowyer et al. 1964), the total energy output of the Crab Nebula from radio to X-rays is an incredible 10^{38} erg s^{-1} or about 100,000 times the energy output of our Sun, even though over 900 years have passed since the supernova explosion! In 1968, the discovery of a pulsar in the nebula by David Staelin of MIT and Edward Reifenstein of the National Radio Astronomy Observatory solved the mystery (Staelin & Reifenstein 1968); the pulsar, rotating 30 times per second, was the powerhouse supplying the tremendous energy radiated by the nebula. The pulses are easily detected at both X-ray and optical wavelengths (Figure 4.4). Figure 4.5 is a movie combining four months of observations to show the rings and jets expanding from the pulsar. Shortly after the association of the Crab pulsar with a supernova event, observers at the University of Sydney identified the Vela pulsar with the Vela Supernova Remnant (Figure 4.6). Today, among over 1000 radio pulsars known, only the Crab and Vela are detected to pulse in optical light (Figure 4.7). This is likely a selection effect because optical light is easily absorbed by the interstellar medium. Vela is the nearest pulsar to Earth and the Crab is above the Galactic plane so both are relatively unobscured.

The Crab Pulsar is slowing down at a rate of about 15 parts in a million each year, so that at the time of its birth it would have been spinning at 60 times a second. Pulsars spinning as fast as hundreds of times per second, called "millisecond pulsars," have magnetic fields a thousand times weaker than the Crab pulsar. Most astronomers believe that the weak-field millisecond pulsars were born rotating slowly and then sped

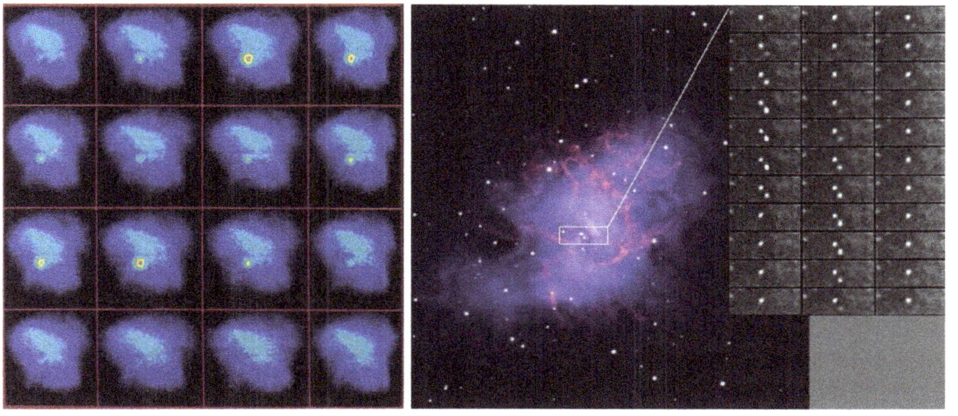

Figure 4.4. Left: Images obtained with the High Resolution Instrument (HRI) on the Einstein X-ray satellite showing the Crab pulsar brightening and fading with each pulse. (Reproduced courtesy of Seward. NASA; Seward & Charles 2010.) **Right:** An image of the Crab Nebula taken with the KPNO 4 m Mayall telescope showing the central pulsar which is slightly below and to the right of the central star turning on and off (the central star is an unrelated object in the foreground). The inset shows 33 images each representing a time slice of about 1 millisecond in the pulsar period. (Reproduced courtesy of N. A. Sharp/AURA/NOAO/NSF, Bill Schoening.)

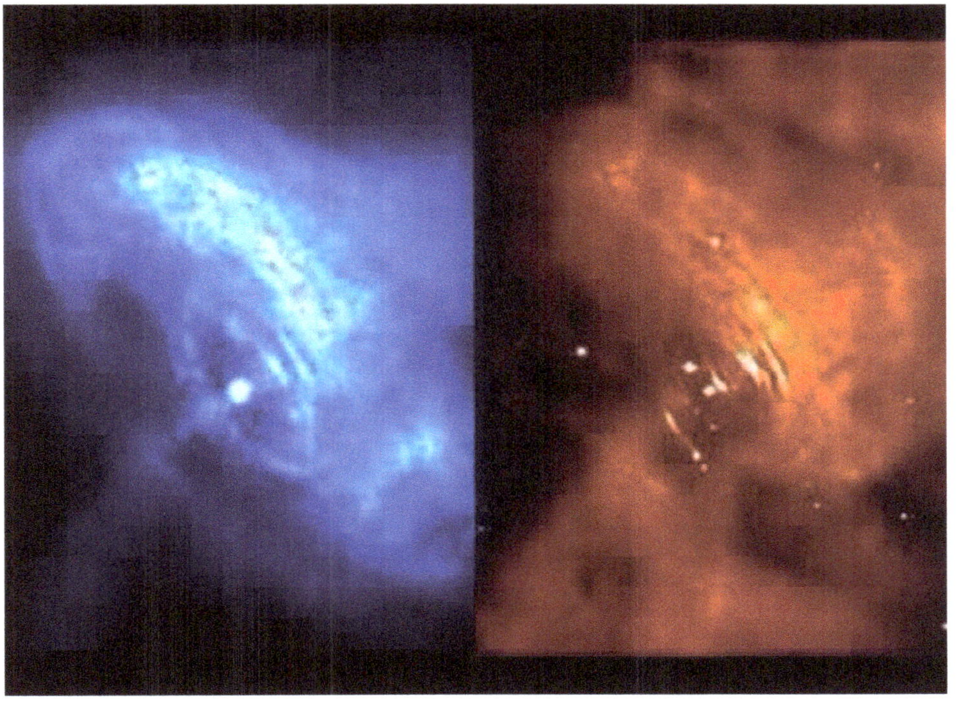

Figure 4.5. An animation combining four months of Chandra X-ray (blue) and Hubble optical (red) images of the Crab pulsar and surrounding area depicts the rings and pulses emerging from the pulsar. (Reproduced courtesy of NASA/CXC/HST/ASU/J. Hester et al.) Animation available at http://doi.org/10.1088/2514-3433/ac81b3.

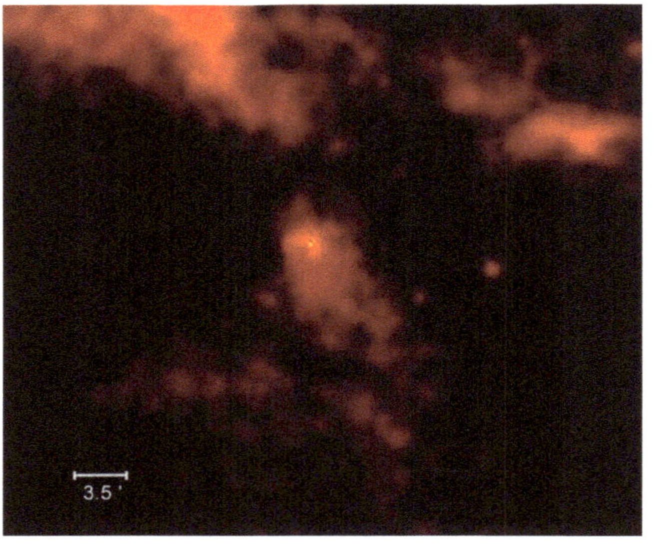

Figure 4.6. This X-ray image from the Chandra observatory shows the central 5.7e13 km (6 light-years) of the Vela supernova remnant with the Vela pulsar which rotates 11 times a second and is located about 7.6e15 km (800 light-years) from Earth. The X-ray rings and jet are very similar to those found around the Crab pulsar. The crossbow shape is roughly 1.9e12 km (0.2 light years) across. The pulsar itself is moving in the direction of the X-ray jet. (Reproduced courtesy of CXC, SAO, NASA, Garmire.)

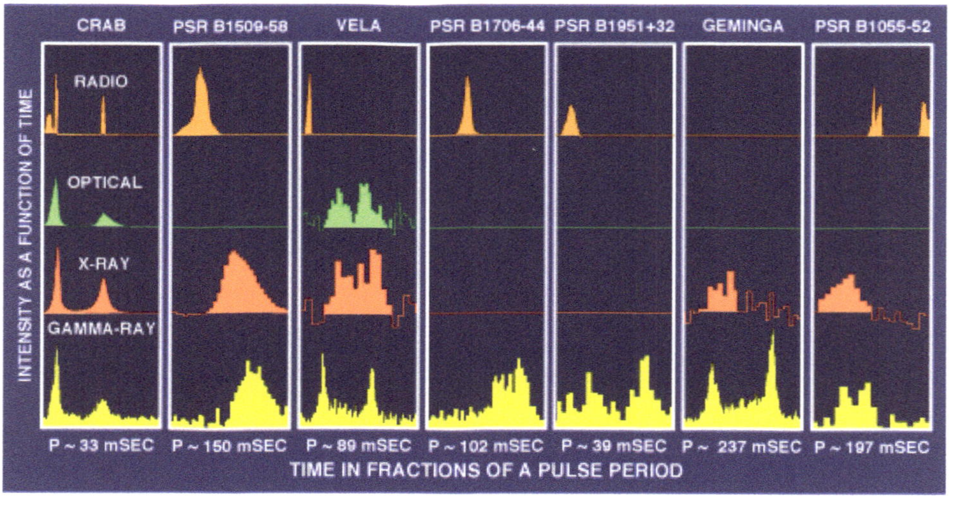

Figure 4.7. The panels in this image illustrate the pulses produced across the electromagnetic spectrum by seven pulsars. Each panel shows the variation of light during one rotation of the pulsar. The period of rotation is given in milliseconds. The Crab and Vela pulsars are the only pulsars detected at optical wavelengths. (Credit: NASA/HEASARC).

up by gaining angular momentum from gaseous material "accreted" from an orbiting stellar companion (see Chapter 5: From Matter to Energy).

In 1998, Frank Marshall of NASA's Goddard Space Flight Center announced the discovery of a neutron star spinning at more than 60 times per second (5 million times as rapidly as the Earth), and calculated that it could have been spinning as fast as 150 times per second or more when it formed some 4000 years ago (Marshal et al. 1998). Most astronomers had not previously believed that pulsars could *form* with such a rapid spin. The newly discovered pulsar, N157B, establishes a link between fast-spinning pulsars with relatively weak magnetic fields and slow-spinning ones with strong fields, suggesting there may be a natural continuum between the two known types.

4.2.1 Binary Pulsars

Because neutron stars are very dense, they have very high moments of inertia (i.e., their rotation is very stable) which makes them good clocks. The 'best' of them are clocks about as good as our best atomic clocks; observers can now predict the arrival times of pulses a year ahead with an accuracy better than a millisecond. The Hulse–Taylor pulsar, PSR 1913 + 16, has another (non-pulsing) neutron star as its companion (Hulse & Taylor 1974). The pulsar is a normal radio pulsar, not a millisecond pulsar, with a rotation period of 59 milliseconds. The two stars are so close that their orbital period is less than eight hours, but no matter streams between them; they interact only by means of their mutual gravitational attraction. The orbit of the pulsar can be described in great detail because the arrival times of its pulses at the Earth are like the ticks of an accurate clock moving very rapidly in a strong gravitational field, which is a situation perfectly suited for a test of Einstein's General Theory of Relativity (Figure 4.8(a)). According to non-relativistic, or Newtonian, dynamical theory, the orbits of both stars should be ellipses with a fixed orientation, and the orbital period should be constant. Measurements of the arrival time of the pulses have shown significant differences from the simple Newtonian orbits. The most obvious is that the orbit precesses by 4.2 degrees per year. There is also a small, but very important, effect on the orbital period, which is now known to be shrinking by 67 nanoseconds (less than one ten-millionth of a second) each orbit. This shrinking orbital period represents a loss of energy, which precisely matches that predicted from gravitational radiation as predicted by Einstein's GR. The observations of PSR 1913 + 16 have provided a convincing demonstration of the existence of gravitational waves and served as a strong incentive to develop detectors for direct detection of gravitational waves.

4.2.2 Gamma-Ray Pulsars

In 1977 one of NASA's Small Astronomy Satellite (SAS-2) detected an excess of gamma-ray emission toward the constellation Gemini (Thompson et al. 1977). The source was confirmed by observations with NASA's COS-B satellite (Masnou et al. 1977). Although the second brightest source to be detected in gamma rays (Figure 4.8(b)), its nature remained a mystery for nearly 20 years and it was aptly

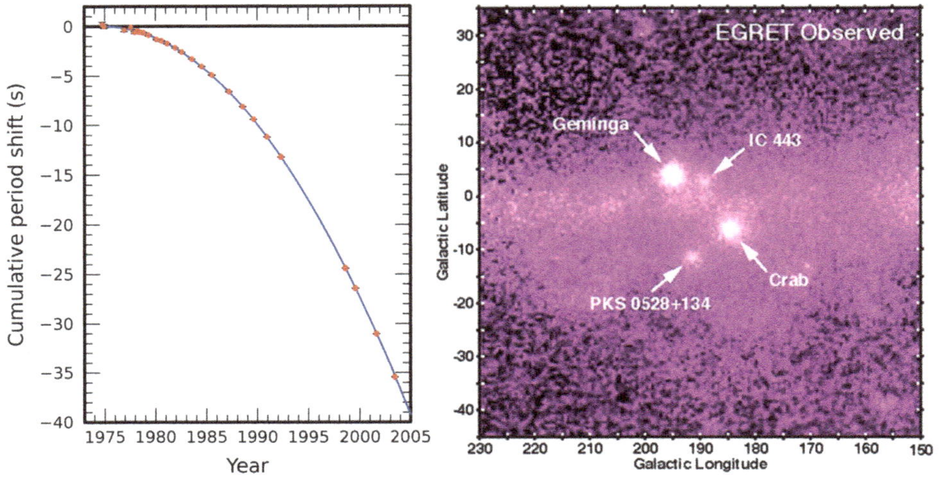

Figure 4.8. Left (a): The decrease of the orbital period of the Hulse–Taylor pulsar is only 40 s over 30 years: this agrees precisely with the predictions for loss due to gravitational waves (the blue curve on the plot) from Einstein's theory of General Relativity. This was the first indirect proof of the existence of gravitational waves. (Reproduced courtesy of Inductiveload.) **Right (b):** This EGRET image shows Geminga (above and left of center) and the Crab Nebula (below and right of center), two bright sources of gamma rays in the direction of the Galactic anticenter. Both Geminga and the Crab contain pulsars that pulse at gamma-ray energies. Strikingly, Geminga mainly pulses in the X-ray and gamma-ray regimes with very little pulsation at other wavelengths, while the Crab pulsar is observed to generate optical and radio pulses as well. The faint source visible just below center is the active galaxy designated PKS 0528 + 134. (Reproduced courtesy of NASA/CGRO.)

named 'Geminga,' a contraction of 'Gemini gamma-ray source' which also happens to mean "it's not there" in Milanese dialect. Higher energy radiation such as gamma rays form much broader beams than at radio wavelengths and can be detected even when the magnetic axis is not aligned toward the Earth (Bignami et al. 1977; Figure 4.3(b)). In 1991, Geminga was finally detected in X-rays by Jules Halpern of Columbia University and Steven Holt of the Goddard Space Flight Center using the German ROSAT satellite (Halpern & Holt 1992). Geminga was pulsing with a period of 0.24 s. Such a period could only be associated with the spin of a neutron star and indeed its gamma-ray pulse shape is very similar to that of the Crab and Vela radio pulsars (Figure 4.7) yet Geminga was not detected in radio waves. Observations with the Chandra X-ray observatory and the Spitzer infrared observatory show long narrow tails that are thought to be narrow jets ejected at the poles and swept back as the pulsar moves supersonically through the Galaxy (Figure 4.9(a)).

In 2004, Halpern and others reported another mysterious X-ray source in the supernova remnant CTA-1 (Halpern et al. 2004). Using Chandra they detected "a point source, a compact nebula, and a bent jet, all of which are characteristic of energetic, rotation-powered pulsars," but deep radio observations with the VLA did not find any radio emission. Finally, in 2008, the Fermi gamma-ray space telescope detected 318.86 millisecond gamma-ray pulses from the point source (Abdo et al. 2008; Figure 4.9(b)). Since then Fermi has detected many such gamma-ray pulsars that are 'dark' in radio waves (Figure 4.10). It is likely that such pulsars radiate optical and radio light in a beam too narrow to be detected at Earth.

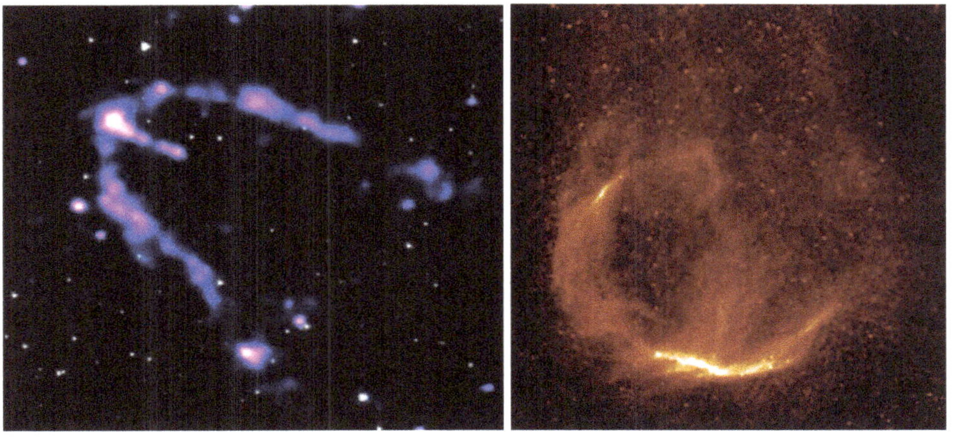

Figure 4.9. Left (a): Geminga observed in X-rays with Chandra (blue and purple) and in infrared with Spitzer (white). The infrared shows background stars. The X-rays (gathered over several years for nearly 8 days on source) show distinct arced trails that are half a light-year long; there is also a narrow structure directly behind the pulsar. (Reproduced courtesy of NASA; X-ray: NASA/CXC/PSU/B. Posselt et al.; Infrared: NASA/JPL-Caltech; Illustration: Nahks TrEhnl.) **Right (b):** The central point source in the CTA-1 supernova remnant has been identified as a gamma-ray pulsar by NASA's Fermi space telescope. (Reproduced courtesy of NASA, S. Pineault (DRAO).)

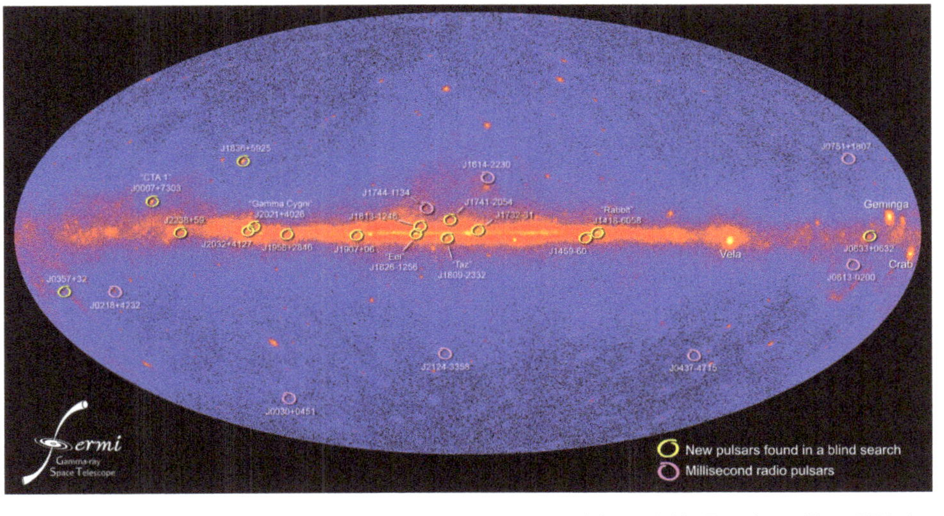

Figure 4.10. Location of pulsars detected first in gamma rays with NASA's Fermi satellite. CTA-1 and Geminga are noted. (Credit: NASA/DOE/Fermi/LAT Collaboration).

4.3 Gamma-Ray Bursts

The first gamma-ray burst was detected in 1969 by the Vela satellites, spacecraft that were deployed by the U.S. Department of Defense to check for violations of the Nuclear Test Ban Treaty of 1963 (Figure 4.11). The Vela satellites detected intense brief bursts of gamma-rays that turned out to be of cosmic origin. A total of 73 GRBs were recorded in the ten year interval July 1969–April 1979. Had the Vela satellites not been needed for security purposes, it is possible that we would still be unaware of GRBs.

When the data were declassified and the first announcement of GRBs reported by Klebesadel, Strong, and Olson (Klebesadel et al. 1973), there was great excitement in the astronomy and physics communities. GRBs are among the brightest sources in the sky producing, for the duration of the burst (which can range anywhere from 30 milliseconds to over 15 minutes) more energy flux than any single object in the sky, other than our SUN, in any wavelength region. For the duration of the burst the source outshines all other sources of cosmic gamma-rays added together. Unlike the X-ray sources discovered less than a decade earlier and very quickly identified with optical sources, no counterparts to GRBs were forthcoming, the main reason being that gamma-ray detectors were not able to pinpoint the bursts accurately enough to search for counterparts. With no information but extremely energetic bursts that appeared to go off in random directions, theories abounded. As early as 1975, Mal Ruderman of Columbia University wrote an article in the Annals of the New York Academy of Science demolishing 21 theories explaining GRBs (at a time when barely 21 bursts had been detected; Ruderman 1975).

In the mid-eighties, two schools of thought formed regarding the nature of GRBs. In one, the GRBs were thought to originate from Galactic neutron stars. X-ray pulsars, which continuously produce gamma-rays through accretion from a companion star, and X-ray bursters, which produce short outbursts of X-rays through thermonuclear

Figure 4.11. Left: Launch of Atlas LV-3A Agena D rocket with Vela 3A (Vela 5), Vela 3B (Vela 6) and ERS-17 satellites, 1965 July 20, Cape Canaveral Air Force Station. One of six pairs sent up by the US Air Force to monitor nuclear tests in space. (Reproduced courtesy of NASA/US Air Force Space and Missile Systems Center.) **Center:** The Vela 5A/B pair in the clean room. (Reproduced courtesy of NASA/Vela Project.) **Right:** A profile of the first gamma-ray burst detected by one of the Vela satellites on 1967 July 2. (Reproduced courtesy of NASA/Vela Project.)

explosions, are two possible sources. The theories combined the physics of both systems in ways that produce short, intense bursts of gamma-rays. This class of theory predicts that GRB sources are distributed in the plane of the Galaxy, because that is where one finds most neutron star systems in our Galaxy (for example, the radio pulsars depicted in Figure 4.2); so a plot of GRB positions on the sky should show a strong concentration of bursts along the Milky Way. A second school held that the bursts are produced in very distant galaxies, billions of light-years away. This implies an uniform spherical distribution since the large-scale universe looks roughly the same in all directions, but requires the sources to produce enormous amounts of energy.

The Burst and Transient Source Experiment (BATSE), an all sky monitor on NASA's Compton Gamma-Ray Observatory (CGRO) launched in 1991, showed conclusively that there was no concentration along the Galactic plane (Figure 4.12). The BATSE observations, which span the energy range 20–600 keV, show that we are at, or very close to, the center of a spherical distribution of burst sources. The extragalactic nature of GRBs was also established by the Large Area Telescope (LAT) on NASA's Fermi Gamma-Ray Observatory launched in 2008 (Figure 4.13). The LAT covered only 20 percent of the sky but it covered energy ranges from 20 MeV to 300 GeV. Today GRBs constitute some of the farthest objects that can be seen with the naked eye (Figure 4.14).

4.3.1 Magnetors, Anomalous X-Ray Pulsars

No persistent, strictly periodic behavior has been seen from GRBs, and of the 2704 observed by BATSE only three appeared to repeat; these have been associated with

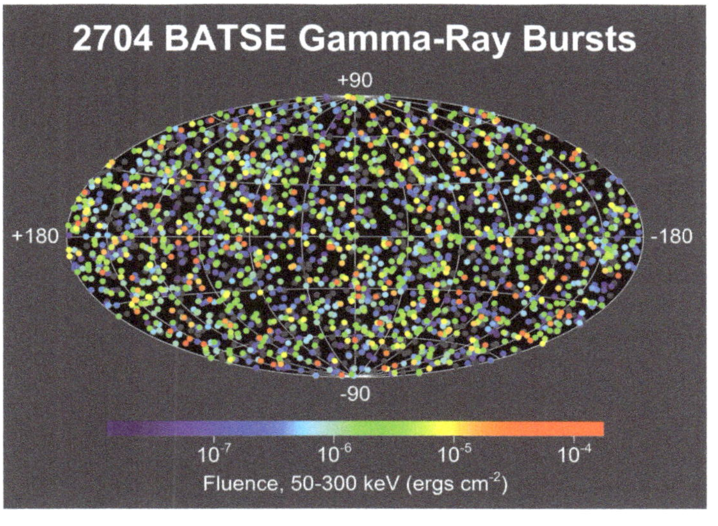

Figure 4.12. From 1991 to 2000, NASA's Burst and Transient Source Experiment (BATSE) detected 2704 GRBs in the energy range 20–600 keV. The final sky map of GRB locations and intensities shows them to occur randomly in every direction with no special symmetry that might be associated with objects within our Galaxy. (Reproduced courtesy of NASA/CGRO.)

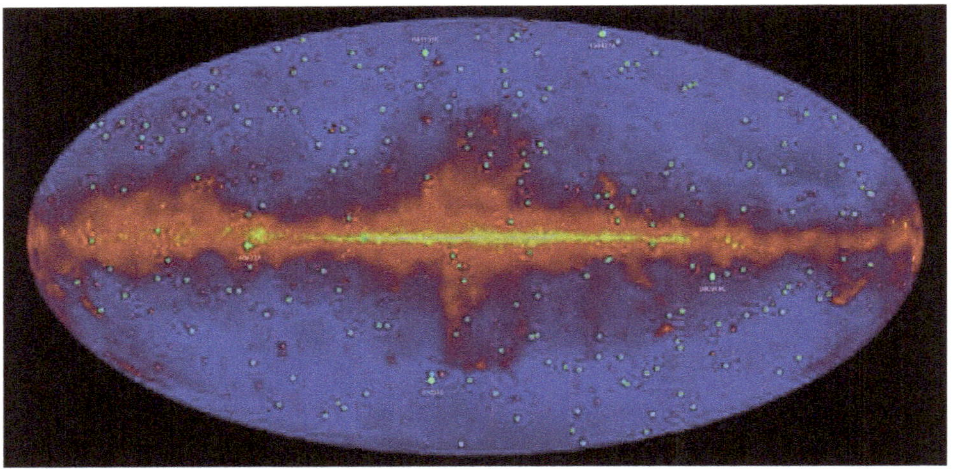

Figure 4.13. Constructed from nine years of data from the LAT on the Fermi gamma-ray telescope, this map shows how the gamma-ray sky appears at energies above 10 GeV. The plane of our Milky Way Galaxy runs along the middle of the plot. Brighter colors indicate brighter gamma-ray sources. Green dots show the locations of 186 GRBs. Some noteworthy bursts are highlighted and labeled. (Reproduced courtesy of NASA/DOE/Fermi LAT Collaboration.)

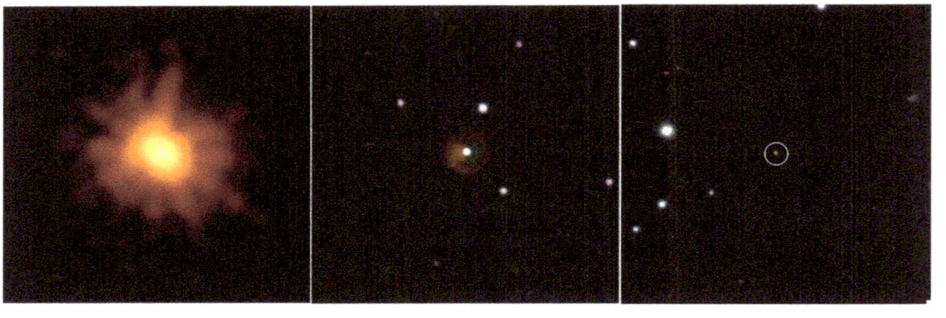

Figure 4.14. Left (a): An X-ray image of GRB080319B is at a distance of 7.5e25 km (7.5 billion light-years). At the moment of its burst it was the most distant object that could be seen by the naked eye and possibly the brightest object ever detected. Its luminosity was over 2.5 million times greater than the brightest known supernova. **Center (b):** An ultraviolet/optical image of GRB8080319B taken with instruments on the Swift Satellite. (Credit: NASA Swift Team, Stefan Immler (GSFC) et al.) **Right (c):** The infrared afterglow of GRB090423, the farthest explosion detected (as of October 2010) was captured by the Gemini North. It has a redshift of 8.2 which corresponds to a distance of over 13 billion light-years. (Credit: Gemini Observatory / NSF / AURA, D. Fox & A. Cucchiara [Penn State U.]), and E. Berger [Harvard]).

magnetars (Kouveliotou et al. 1998). Magnetars are neutron stars with magnetic fields up to a 1000 times greater than normal pulsars (from 10^{13} to 10^{15} G). Anomalous X-ray pulsars (van Paradijs et al. 1995), called 'anomalous' because they did not share characteristics with either classical radio pulsars (powered by rotation), or X-ray binary pulsars (powered by accretion, see Chapter 5) also turned out to be magnetars (Woods & Thompson 2006).

4.3.2 Short and Long Gamma-Ray-Bursts: SGRBs and LGRBs

A striking feature of the time profiles of GRBs is their diversity (Figure 4.15(a)). Some burst light curves are spiky with large fluctuations on all time scales, while others show rather simple structures with few peaks. Sometimes both characteristics are present within the same burst at different times in its evolution. BATSE also determined that GRBs separate into two distinct classes (Kouveliotou et al. 1993; Figure 4.15(b)); those with bursts that lasted shorter than 2 s (short GRBs or SGRBs) and those that lasted longer (LGRBs). SGRBs have larger percentages of higher energy gamma rays than LGRBS and can release more energy in one second than our Galaxy emits over one year (Rezzolla et al. 2011).

In the early 1990s Peter Meszaros of Pennsylvania State University and Martin Rees at the Institute of Astronomy predicted that interaction of the gamma-rays with the surrounding medium should result in emission at longer wavelengths (and over longer timescales) from X-rays to radio. They called this a 'fireball model': a blast from an explosion running into the surrounding interstellar gas at nearly the speed of light (Meszaros & Rees 1993, 1997). This model led to extensive searches for "afterglows." BATSE detected on average one burst per day, but it did not have the spatial resolution to pinpoint the position of the object or objects causing the burst to a small region of the sky, making searches at other wavelengths difficult.

4.3.3 LGRB Counterparts—X-Rays, Optical, Radio

A breakthrough for LGRBs came in 1997 with the launch of an Italian–Dutch satellite, BeppoSAX. BeppoSAX had both X-ray and gamma-ray detectors

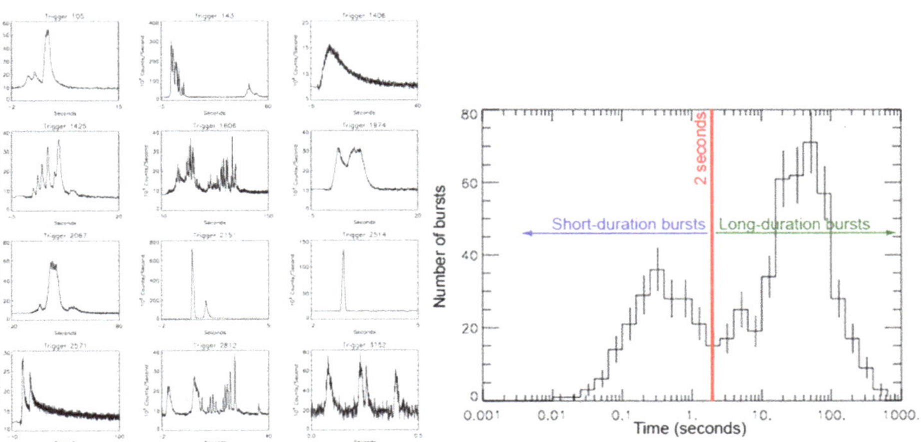

Figure 4.15. Left: Light curves (plots of the intensity of emission as a function of time) of 12 bright GRBs detected by the BATSE instrument on NASA's Compton Gamma-Ray Observatory showing the diversity of behavior. Current understanding equates short period (less than 1 s) and long period (greater than 2 s) GRBs with different progenitors. (Reproduced courtesy of NASA/GSFC/Bonnell.) **Right:** Graph of the time versus number of bursts for the GRBs observed by the BATSE instrument on the Compton Gamma-ray Telescope. There appears to be two distinct types: shorter or longer than 2 s in duration. (Reproduced courtesy of NASA/CGRO.)

(Costa et al. 1997). The X-ray detector had much better positional resolution than the gamma-ray detector and the X-ray position was accurate enough to pinpoint the location of the source. Equally important for the detection of counterparts, BeppoSAX's "turn-around time" (the time from when a detection is made, a position determined, and that position published to the world) was short enough that optical and radio telescopes had a chance to find counterparts before they faded.[6] GRB 970228, with a burst that lasted 80 s, was detected in X-rays within hours of its initial discovery in gamma-rays (Figure 4.16).[7] The X-ray, optical, and radio emissions of the afterglow have been essential in revealing details of the fireball created by the initial explosion (Galama et al. 1997; Tavani 1997).

The accurate X-ray position obtained by BeppoSAX led to the detection of an optical counterpart as reported by Paul Groot and collaborators from the

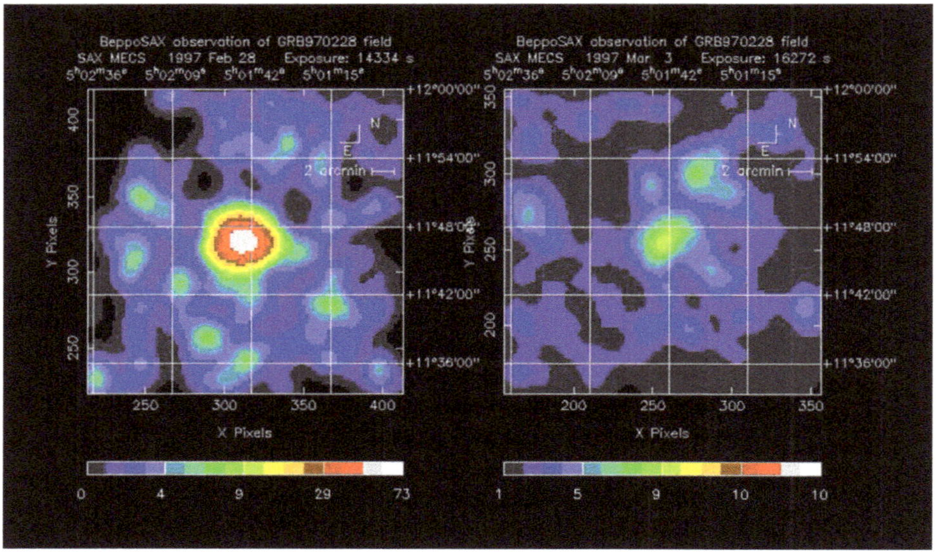

Figure 4.16. GRB 970228 was the first GRB that was detected almost immediately in X-rays with the Gamma-Ray Burst Monitor (GRBM) and one of the Wide Field Cameras (WFCs) on the Dutch-Italian BeppoSAX satellite on 1997 February 28. The burst lasted about 80 s and had multiple peaks. Three days later on 1997 March 3, the X-ray source had faded. (Reproduced courtesy of NASA/BeppoSAX Data Center (SDC) and the Agenzia Spaziale Italiana (ASI).)

[6] In 1988, the Japanese satellite Ginga had the capability to capture both gamma-rays and X-rays; although the X-ray position was better determined than the gamma-ray position, the error-box was still large, and by the time the position was determined and published to the world (the record was 18 days) the X-ray counterpart had already faded and could not be determined.

[7] Since the positions of GRBs are hard to pinpoint, the naming convention has been simply the year, month, and day of the burst detection. This convention has worked so far, because BATSE, which covered essentially the whole sky, detected roughly one burst per day during its lifetime (1991–2000). NASA's SWIFT satellite launched in 2004 primarily for the study of GRBs finds roughly 100 per year.

Astronomical Institute Anton Pannekoek in March 1997 (Groot et al. 1998). Comparing observations taken with two ground-based optical telescopes, the William Herschel Telescope on 1997 February 28 and the Isaac Newton Telescope on 1997 March 8, Groot found an optical object that flared briefly and then faded away. Observations of this position by HST gave a hint of extended emission from a possible host galaxy (Figure 4.17); the source of the burst is clearly offset from the center of the host galaxy.

The second optical counterpart of a LGRB was found on 1997 May 8 (Sahu 1997). In this case, Jules Halpern, then at Berkeley, alerted Howard Bond of Space Telescope Science Institute, who was observing at the Kitt Peak National Observatory. Although using a small telescope (0.9 m in diameter), Bond was able to obtain the first optical CCD frame less than 6 hours after the burst occurred (Bond 1997). An image obtained on the second night showed a source in the center of the error circle that had brightened by about one magnitude (GRB970508; Figure 4.18). Even more exciting, spectra obtained with the KECK II (at 10 m diameter, the world's most powerful ground-based optical telescope) showed absorption lines with a redshift that corresponded to several billion light-years

Figure 4.17. A Hubble Space Telescope image of the long-duration gamma-ray burst GRB 970228. The arrow indicates the GRB and the extended emission to the lower right of center is interpreted to be the host galaxy within which the GRB is embedded. (Reproduced courtesy of NASA/Andrew Fruchter (STScI), Elena Pian (ITSRE-CNR), NASA/ESA.)

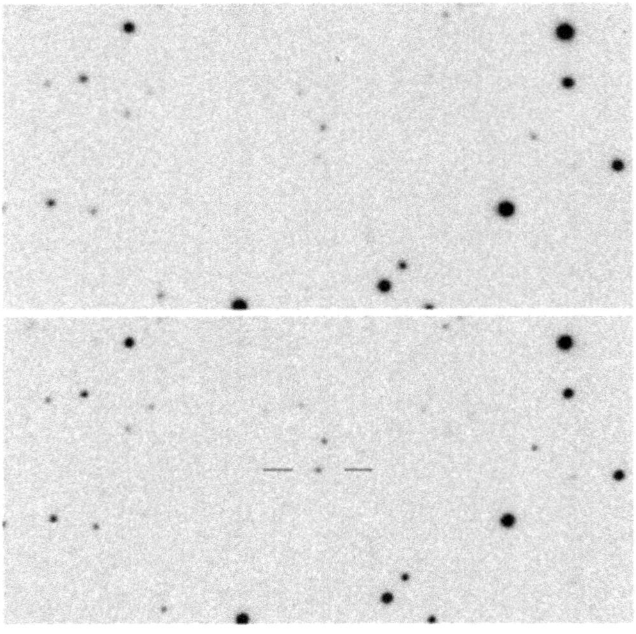

Figure 4.18. Discovery images for the optical counterpart of GRB 970508, obtained with the KPNO 0.9 m telescope by H. Bond. The lower panel shows an image taken on 1997 May 9. The upper panel shows the same sky region imaged on 1997 May 10. The optical counterpart (marked with dashes) brightened by more than a factor of two from 9 May. (Reproduced courtesy of Sahu et al.)

from Earth (over half the way across the observable universe), establishing its extragalactic nature (Bloom et al. 1998). This implies that for a few seconds the burst was over a million times brighter than an entire galaxy.

GRB970508 was also the first GRB for which a radio counterpart was obtained using the Very Large Array (VLA). Dale Frail and collaborators at the National Radio Astronomy Observatory in Socorro, New Mexico observed GRB970508 less than four hours after its discovery, but the radio emission was not detected until May 13 (unlike the X-ray and optical counterparts which flared at the same time or very shortly after the gamma-ray burst itself; Frail et al. 1997). For about three months the radio emission rose and fell several times, brightening by factors of two or three, and then the fluctuations ceased and the object appeared to remain at a steady level of brightness. Several origins have been suggested for the unexpected variability in the radio output. One idea is that we are seeing the same scintillation (the radio analogy of stars twinkling in the night sky) that Hewish used to search for radio quasars. If scintillation were the explanation, then the object would have to be very small. Another idea is that the variations in the radio emission are not connected to the burst itself, but are arising within an active galaxy in which the burst occurred.

Observations with the VLBI, a continent-wide radio telescope system that shows the object to be less than a few light-years in size, appears to favor the former interpretation (Taylor et al. 1997). The VLBI observations show a level of detail more than 50 times greater than that of the Hubble Space Telescope at optical wavelengths. The VLBI observations enabled astronomers to pinpoint a position for the 1997 May 8 burst to within a thousandth of a second of arc. The VLBI observations also showed that the object is expanding at speeds very close to the speed of light (as determined by its angular movement across the sky). Its size three months after the initial GRB was found to be about one-tenth of a light-year, or 170 times the distance between the Sun and Pluto. The cessation of fluctuations would be due to the source becoming larger than the irregularities in the thin material between the stars: i.e., the same reason why stars which are point like appear to twinkle whereas planets presenting a bigger surface do not twinkle.

4.3.4 Host Galaxies

Observations of regions around GRBs confirmed that they were often associated with galaxies but not coincident with the galactic centers; hence the bursts were not associated with the massive black holes at the centers of galaxies. LGRBs were found to be associated with the brightest, star-forming regions of galaxies, where the most massive stars are found (Figures 4.19; Bloom et al. 2002). SGRBs are found in

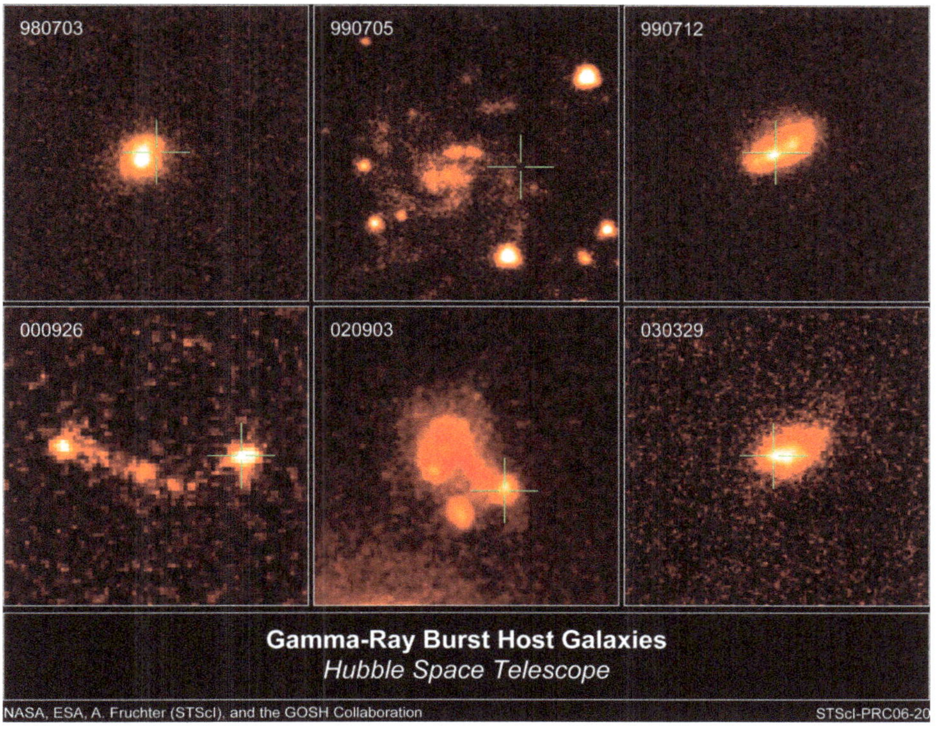

Figure 4.19. Examples of host galaxies of long-duration GRBS as imaged by HST. These tend to occur in the brightest star-forming regions containing massive stars within small, irregular galaxies. The green crosses mark the positions of the GRBs. (Credit: NASA, ESA, Andrew Fruchter (STScI), and the GOSH Collaboration).

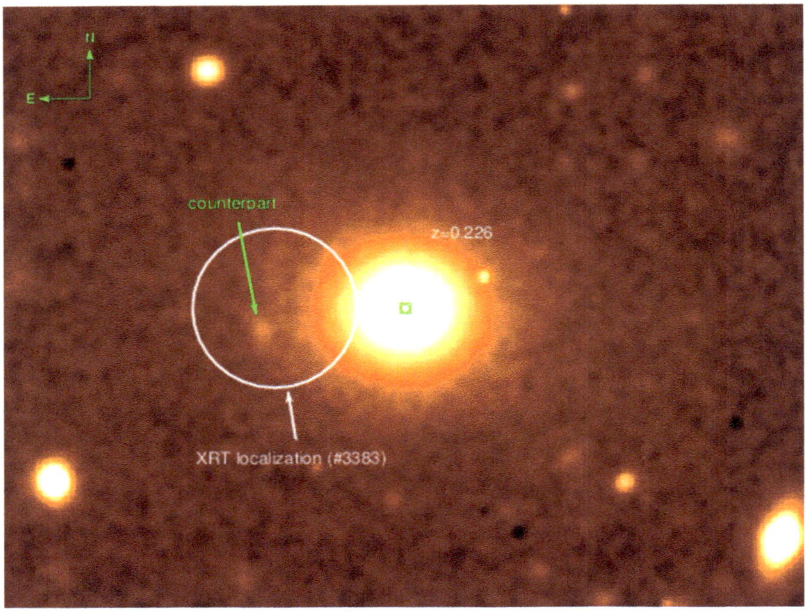

Figure 4.20. The host galaxy associated with the SGRB, GRB050509b, is a giant elliptical at a redshift of $z = 0.226$. The error circle from the BeppoSAX XRT is shown in white on this WIYN 600s Sloan image. (Reproduced courtesy of J. Bloom.)

older galaxies, such as giant ellipticals, where star-formation has essentially ceased (Figure 4.20; Bloom et al. 2002b; Prochaska et al. 2006).

4.3.5 Collapsars, Hypernovae

LGRBs also appeared to be associated with supernovae; the GRB counterpart pictured in Figure 4.21 did not simply fade away. Follow-up observations with Hubble, the Polish Optical Gravitational Lensing Experiment (OGLE), and one of the Magellan telescopes in Chile showed that it brightened again days after the burst (Bloom et al. 2002a). This secondary brightening is characteristic of a Type Ic supernova event and the timing is consistent with a supernova explosion at the time of the GRB. The X-ray, optical, and radio behaviors of LGRBs are all consistent with the explosions of massive stars, sometimes referred to as collapsars or hypernovae (Figure 4.22; MacFadyen & Woosley 1999).

4.3.6 SGRB Counterparts, Mergers, Kilonova, r-Process Elements

The first counterpart for SGRBs was obtained nearly eight years after the BeppoSAX success with LGRBs. GRB 050509B was observed in 2005 with the X-ray telescope on NASA's Swift observatory (Gehrels et al. 2005). Within 50 s of localizing the X-ray position the Swift Ultraviolet/Optical Telescope (UVOT) took a 100 s exposure in the V band (Figure 4.20) and found a luminous giant elliptical

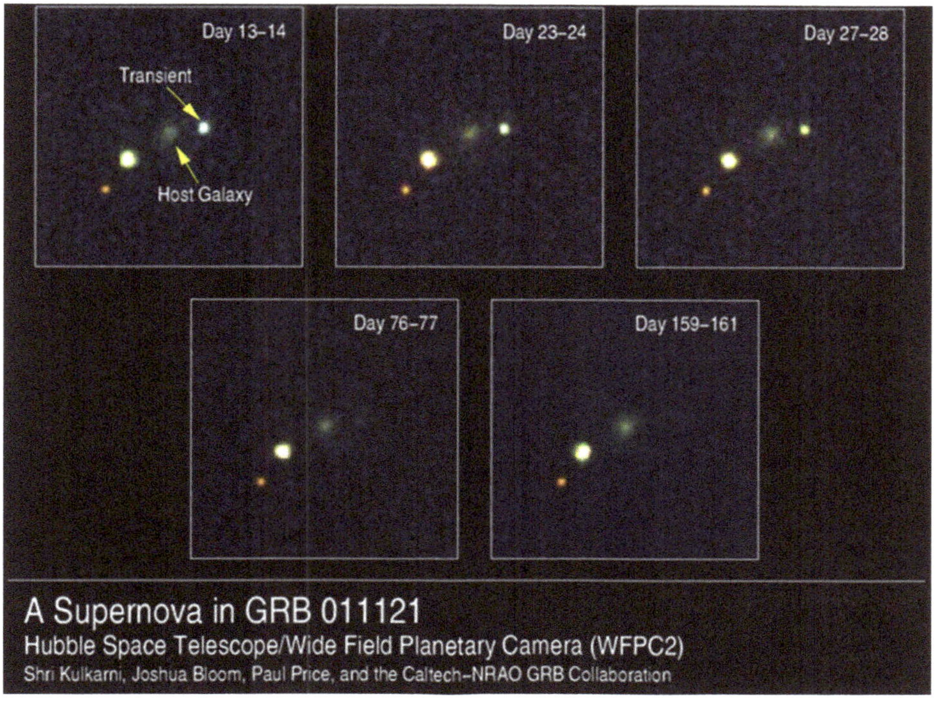

Figure 4.21. A gamma ray burst detected on 2001 November 21 (GRB011121) is shown in this series of Hubble Space Telescope images recorded from 2001 December 4 to 2002 May 5 (13 through 161 days after the burst). The fuzzy glow to the left of the burst is a galaxy within which the burst likely occurred. The red and white blobs to the left of the galaxy are foreground stars within our own Milky Way. (Reproduced courtesy of Kulkarni, Bloom, & Price, Caltech, NRAO.)

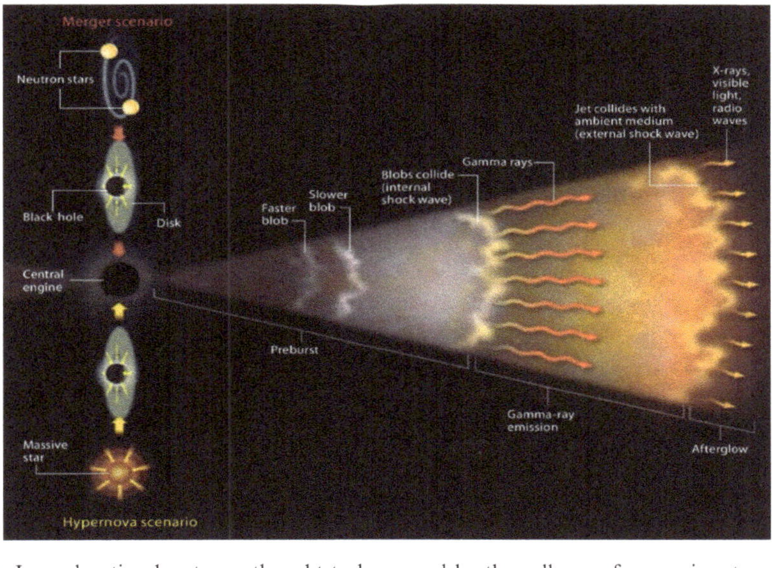

Figure 4.22. Long-duration bursts are thought to be caused by the collapse of a massive star generating a hypernova; short duration bursts are thought to be caused by neutron star–neutron star or neutron star–black hole mergers. Both end up as black holes that produce GRBs and X-ray, optical, and radio afterglows. (Reproduced courtesy of NASA.)

galaxy at a redshift of 0.226, somewhat offset from the XRT error circle. In 2013 an SGRB, GRB 130603B. with a burst duration of 0.18s was detected by Swift's Burst Alert Telescope and the Konus instrument (Melandri et al. 2013; Golenetskii et al. 2013). The optical afterglow was identified by the William Herschel Telescope (Levan et al. 2013). Tanvir et al. (2013) obtained HST imaging 9 and 30 days following the burst dubbed it a kilonova (Figure 4.23). The ultraviolet, optical, and infrared emission of a kilonova is thought to be powered by radioactive decay of nuclei produced via r-process nucleosynthesis. X-ray, gamma-ray, and radio the emission is thought to come from jets typical of SGRBs. Berger et al. (2013) found excess infrared emission from GRB 130603B consistent with that of r-process elements that power kilonovae. In 2017, the detection of both electromagnetic radiation and gravitational waves from GW170817, with a binary neutron star merger as a cause of the gravitational waves detected by LIGO, provided the first direct evidence on the origin of SGRBs (Abbott et al. 2017a, 2017b; Evans et al. 2017; Figures 4.24 and 4.25). In addition spectroscopic evidence was obtained for r-process nucleosynthesis from GW170817 (Chornock et al. 2017; Tanvir et al. 2017).

Figure 4.23. A composite image of the host galaxy for the short duration gamma-ray burst, GRB 130603B. Hubble's Wide Field Camera 3 on 2013 June 13, shows a glow in near-infrared light at the source of the GRB, shown in the top right frame. By July 3, the source had faded. This was the first SGRB to be associated with a kilonova: a faint transient created by decay of r-process elements generated in the burst (Tanvir et al. 2013). (Courtesy of NASA, ESA, N. Tanvir (University of Leicester), A. Fruchter (STScI), and A. Levan (University of Warwick) Licensing).

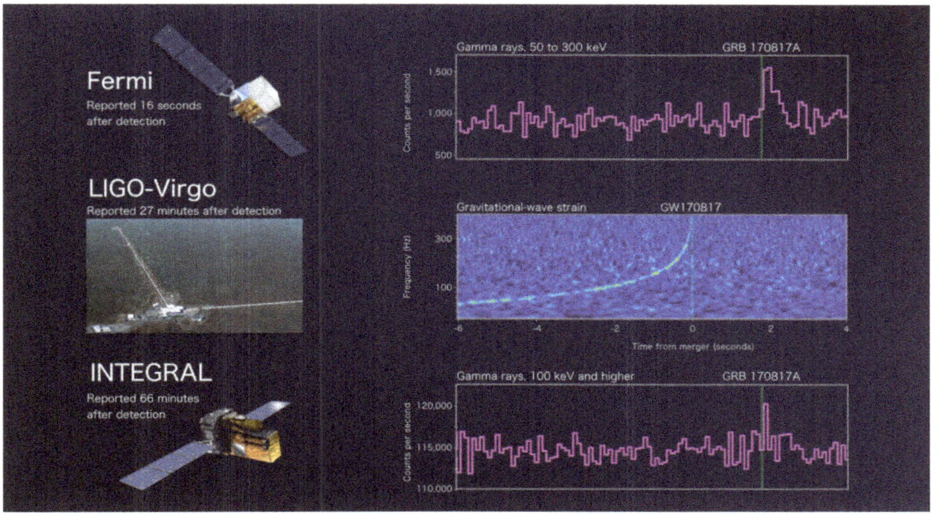

Figure 4.24. The gamma-ray burst (GRB 170817A) detected by Fermi was associated with a LIGO-Virgo event, GW170817, and was also detected by INTEGRAL. (Reproduced courtesy of NASA/GSFC.)

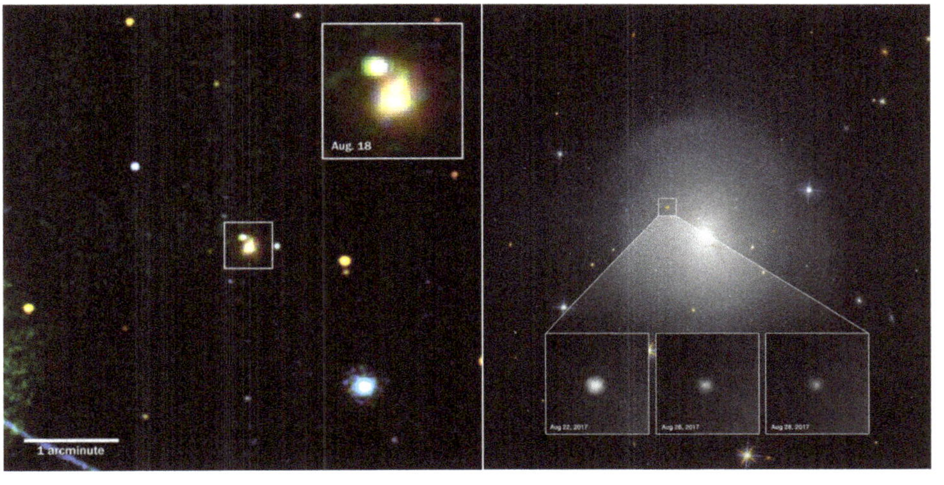

Figure 4.25. Left (a): Fermi Swift discovered GW170817 in ultraviolet light a day after the gravitational wave event. Surprisingly no X-rays were detected. (Reproduced courtesy of NASA.) **Right (b):** Hubble images showed the kilonova clearly on the outskirts of a lenticular galaxy, NGC 4993. This was the first optical counterpart of a gravitational wave event observed. (Reproduced courtesy NASA / ESA / A.J. Levan, University of Warwick / N.R. Tanvir, University of Leicester / A. Fruchter & O. Fox, STScI.)

The fact that SGRBs were found on the outskirts of galaxies, did not appear to be associated with supernovae, and had fainter afterglows, are all consistent with the merger of compact objects (Berger 2014; Figure 4.26). Narayan et al. (1992) had suggested that compact object mergers that resulted in SGRBs should also produce

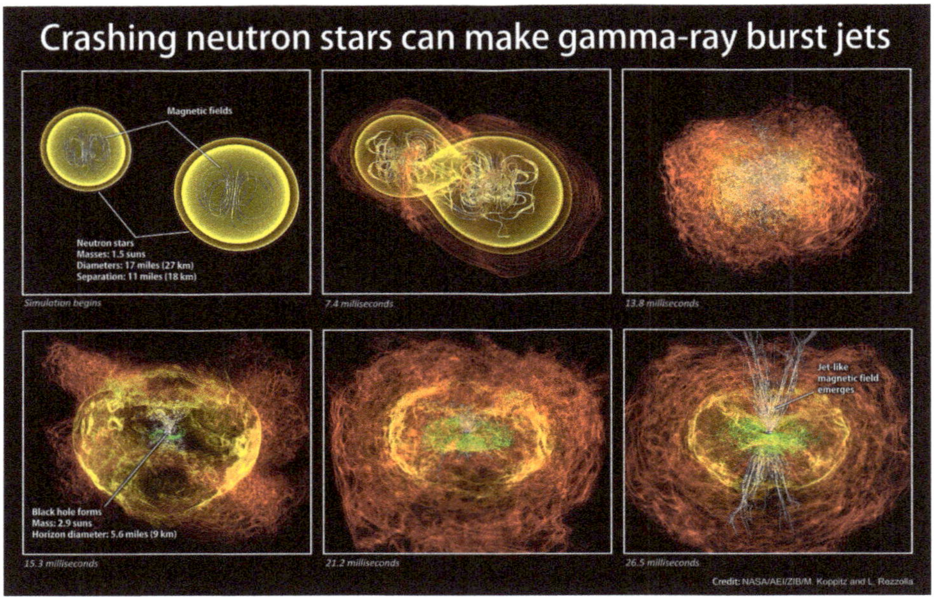

Figure 4.26. Short duration bursts are associated with the collision of two neutron stars or a neutron star and a black hole. Depicted here are snapshots from the animation showing the merger of two neutron stars resulting in a short duration gamma-ray burst. The simulation ran for seven weeks on the Damiana computer cluster at the Albert Einstein Institute (AEI) in Potsdam, Germany. It covers 35 milliseconds of actual time. (Reproduced courtesy of NASA/STScI.) Animation available at http://doi.org/10.1088/2514-3433/ac81b3.

gravitational waves in the frequency range detectable by LIGO (Laser Interferometer Gravitational-Wave Observatory). Since 2017 searches for kilonovae and binary star mergers have become the hottest pursuits in astronomy.

If SGRBs do turn out to be powered by neutron star collisions, then the majority of the new cosmic characters, radio pulsars, anomalous X-ray pulsars, X-ray binaries (see Chapter 5: From Matter to Energy), GRBs, and soft gamma-ray repeaters that were detected first in the wavelength that gave them their names will turn out to be associated with neutron stars: in isolation, highly magnetized, with a supportive companion, or in a highly destructive interaction.

References

Abbott, B. P., Abbott, R., Abbott, T. D., et al. 2017a, ApJL, 850, L40
Abbott, B. P., Abbott, R., Abbott, T. D., et al. 2017b, ApJL, 850, L39
Abdo, A. A., Ackermann, M., Atwood, W. B., et al. 2008, Sci, 322, 1218
Baade, W., & Zwicky, F. 1934, PNAS, 20, 254
Bell Burnell, S. J. 1977, NYASA, 302, 685
Berger, E., Fong, W., & Chornock, R. 2013, ApJ, 774, L23
Berger, E. 2014, ARA&A, 52, 43
Bignami, G. F., Gavazzi, G., & Harten, R. H. 1977, A&A, 54, 951
Bloom, J. S., Djorgovski, S. G., Kulkarni, S. R., & Frail, D. A. 1998, ApJ, 507, L25

Bloom, J. S., Kulkarni, S. R., Price, P. A., et al. 2002a, ApJ, 572, 45
Bloom, J. S., Kulkarni, S. R., & Djorgovski, S. G. 2002b, AJ, 123, 1111
Bond, H. E. 1997, IAU Circ. No. 6654, 1
Bowyer, S., Byram, E. T., Chubb, T. A., & Friedman, H. 1964, Natur, 201, 1307
Chandrasekhar, S. 1931, ApJ, 74, 81
Chiu, H.-Y. 1964, PhT, 17, 21
Chornock, R., Berger, E., Kasen, D., et al. 2017, ApJ, 848, 19
Costa, E., Frontera, F., Heise, J., et al. 1997, Natur, 387, 783
Evans, P. A., Cenko, S. B., Kennea, J. A., et al. 2017, Sci, 358, 1565
Frail, D. A., Kulkarni, S. R., Nicastro, L., Feroci, M., & Taylor, G. B. 1997, Natur, 389, 261
Galama, T., Groot, P. J., van Paradijs, J., et al. 1997, Natur, 387, 479
Gehrels, N., Sarazin, C. L., O'Brien, P. T., et al. 2005, Natur, 437, 851
Groot, P. J., Galama, T. J., van Paradijs, J., et al. 1998, ApJ, 493, 27
Golenetskii, S., Aptekar, R., Frederiks, D., et al. 2013, GCN, 14771, 1
Halpern, J. P., Gotthelf, E. V., Camilo, F., Helfand, D. J., & Ransom, S. M. 2004, ApJ, 612, 398
Halpern, J. P., & Holt, S. S. 1992, Natur, 357, 222
Hewish, A., Bell, S. J., Pilkington, J. D. H., Scott, P. F., & Collins, R. A. 1968, Natur, 217, 709
Hulse, R. A., & Taylor, J. H. 1974, ApJ, 191, L59
Klebesadel, R. W., Strong, I. B., & Olson, R. A. 1973, ApJ, 182, 85
Kouveliotou, C., Dieters, S., Strohmayer, T., et al. 1998, Natur, 393, 235
Kouveliotou, C., Meegan, C. A., Fishman, G. J., et al. 1993, ApJ, 413, L101
MacFadyen, A. I., & Woosley, S. E. 1999, ApJ, 524, 262
Marshall, F. E., Gotthelf, E. V., Zhang, W., Middleditch, J., & Wang, Q. D. 1998, ApJ, 499, 179
Masnou, J. L., Bennett, K., Bignami, G. F., et al. 1977, Proc. 12th ESLAB Symp., ed. R. D. Wills, & B. Battrick, 33
Melandri, A., de Pasquale, M., Barthelmy, S. D., et al. 2013, 2013GCNR, 442, 1
Meszaros, P., & Rees, M. 1993, ApJ, 418, L59
Meszaros, P., & Rees, M. 1997, ApJ, 476, 232
Narayan, R., Paczynski, B., & Piran, T. 1992, ApJ, 395, L83
Prochaska, J. X., Bloom, J. S., Chen, H.-W., et al. 2006, ApJ, 642, 989
Rezzolla, L., Giacomazzo, B., Baiotti, L., et al. 2011, ApJ, 732, L6
Ruderman, M. 1975, NYASA, 262, 164
Sahu, K. C., Livio, M., Petro, L., et al. 1997, ApJ, 489, L127
Seward, F. D., & Charles, P. A. 2010, Exploring the X-ray Universe (Cambridge: Cambridge Univ. Press)
Staelin, D. H., & Reifenstein, E. C. III 1968, Sci, 162, 1481
Tanvir, N. R., Levan, A. J., Fruchter, A. S., et al. 2013, Natur, 500, 547
Tanvir, N. R., Levan, A. J., González-Fernández, C., et al. 2017, ApJ, 848, L27
Tavani, M. 1997, ApJ, 483, L87
Taylor, G. B., Frail, D. A., Beasley, A. J., & Kulkarni, S. R. 1997, Natur, 389, 263
Thompson, D. J., Fichtel, C. E., Hartman, R. C., & Lamb, R. C. 1977, ApJ, 213, 252
van Paradijs, J., Taam, R. E., & van den Heuvel, E. P. J. 1995, A&A, 299, 41
Woods, P. M., & Thompson, C. 2006, in Compact Stellar X-ray Sources (Cambridge: Cambridge Univ. Press)

Chapter 5

From Matter to Energy

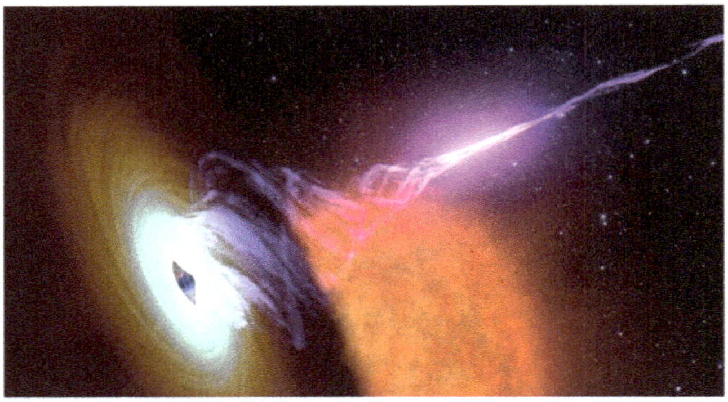

An artist's conception of a black hole with accretion disk and jet. Reproduced courtesy of NASA.

5.1 Introduction

Our most familiar source of light, the Sun, shines because deep in its interior it is converting matter to energy through nuclear fusion. The pressure and temperature in the center of the Sun (due to its own gravity) is such that the hydrogen atoms of which it is mainly composed lose their electrons and the free protons can overcome their mutual repulsion and combine to form helium. The helium atom weighs less than the four hydrogen atoms that comprise it and that small difference, less than one percent of the total mass, is converted to energy. But there are far more efficient means of converting matter to energy in the universe. The most important of these is simply dropping matter onto extremely dense objects such as neutron stars or black holes (a process called accretion). The gravitational potential energy released through accretion is the most efficient means we know of "converting" matter to energy, accretion onto a neutron star has up to 10 times the efficiency of nuclear

doi:10.1088/2514-3433/ac81b3ch5

burning, and accretion onto a black hole is 30–40 times more efficient. Accretion-powered objects range from X-ray binaries (XRBs), which are the most luminous sources in our own Galaxy, to the active galactic nuclei and quasars that are some of the most intrinsically luminous cosmic sources now known.

5.2 Our Sun: The Most Luminous Object in Our Sky

Although a fairly average star, our Sun appears as the brightest object at all wavelengths because it is by far the nearest star to us; the next closest star is 3.8×10^{13} km (four light-years) away. The Sun's energy emission peaks at visible wavelengths, and not surprisingly that is what our eyes are sensitive to. The bright yellow disk that we see in the sky is actually a very thin layer of the Sun, the photosphere. The photosphere emits optical light, but the Sun is actually emitting radiation in every part of the electromagnetic spectrum, as well as electrons and protons (in a solar wind) and neutrinos from its center.

The presence of spots on the photosphere of the Sun were recorded as early as 4th century and they have been continuously monitored since the early 1600s. Both Galileo Galilei (1564–1642) and Christian Scheiner (1575–1650) claimed to have made the first telescopic observations of sunspots and they both noted that the spots appeared to move around the Sun in a regular way (Figure 5.1; Galileo 1613; Scheiner 1630). Scheiner suggested that the spots were the shadows of objects orbiting the Sun, whereas Galileo noted that their changing nature precluded a solid object and claimed that they were due to changes on the surface of the Sun. Galileo's interpretation meant that the Sun itself was rotating, contradicting Aristotle's 4th century claim that the heavens were unchanging. By the 19th century a sunspot cycle

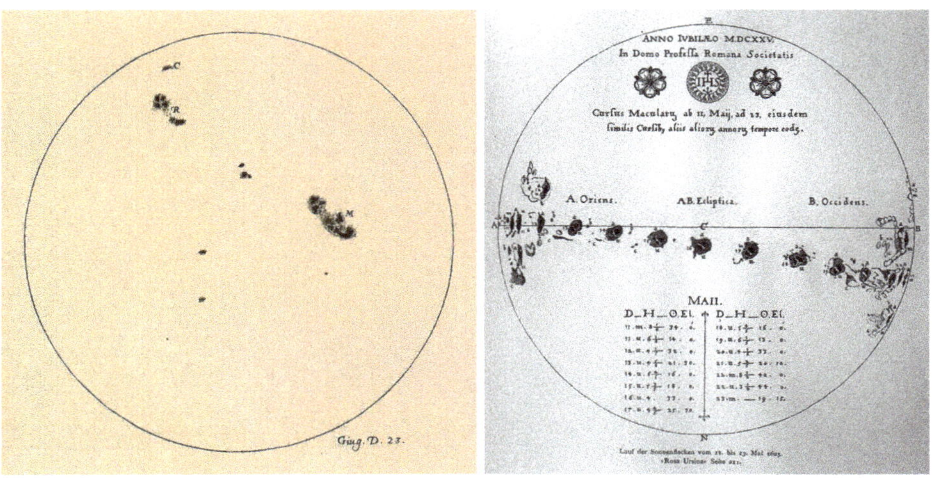

Figure 5.1. **Left:** Galileo's 1612 drawing of sunspots. The spots are shown to be irregular and to change with time. (Reproduced courtesy of Owen Gingerich: first edition of Istoria e Dimostrazioni.) **Right:** Scheiner's sunspot observations of 1611. Scheiner's drawings show a circular spot that stays constant in time. His interpretation is that these are due to objects separate from the Sun and in orbit around it. (Christoph Scheiner, 1575–1650).

with a period of 11 years during which sunspots systematically increased (Solar maximum) and then decreased (Solar minimum) in number was established (Schwabe 1843). In 1908 using a spectroheliograph of his own design, George Ellery Hale (1858–1838; Hale 1908a, 1908b, 1908c) showed that the sunspots were associated with magnetic fields.

Solar activity beyond sunspots is manifest on many levels and on many scales: solar flares, solar prominences, coronal mass ejections, and coronal holes can all be linked to the Sun's magnetic fields, and produce emission from radio waves to gamma-rays. Solar flares reach temperatures of 10 million degrees Kelvin and last up to 10s of minutes. Solar prominences reach temperatures up to 1 million degrees Kelvin, span sizes up to 800,000 km, can develop over one day and last for weeks to months (Figure 5.2). Some prominences result in coronal mass ejections that eject billions of tons of coronal matter and magnetic fields. The speeds of coronal mass ejections can vary from 20 to 3000 km s^{-1}. The matter it ejects can take between half-a-day to three months to reach Earth. Coronal holes are areas of the Sun's corona that are cooler and denser than the surrounding areas and show up as dark regions in X-ray and ultraviolet images. In addition Earth receives neutrinos generated by fusion in the Sun's center and actual ionized particles in a Solar wind from the corona (See Chapter 1, Figure 1.19(b) for an image of the Sun in neutrinos).

The Sun was the first cosmic X-ray source detected. V2 rocket flights conducted by the Naval Research Laboratories in 1948 under the direction of Herbert Friedman showed the Sun to be a powerful source of ultraviolet and X-ray radiation (Friedman et al. 1951). Physical particles, neutrinos, long predicted to be generated by the nuclear processes in the center of the Sun, were predicted by John Bahcall and

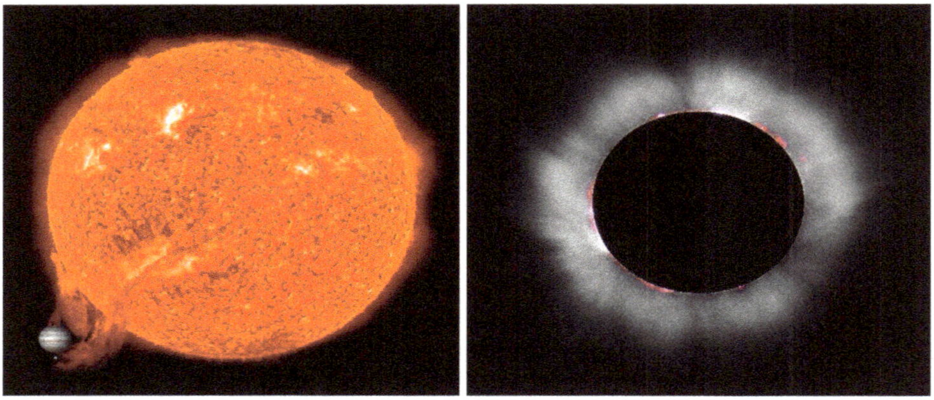

Figure 5.2. Left (a): A solar prominence observed by the Solar Dynamics Observatory (SDO) on 2010 December 6 using a filter for He II 304angstroms. Images of Jupiter and Earth are displayed to show scale. Several sunspots are visible as well as multiple spicules visible on the rim in the chromosphere. (Reproduced courtesy of NASA.) **Right (b):** Photograph of a total Solar eclipse taken in France in 1999. In addition to the corona which becomes prominent during eclipse, several Solar prominences are visible around the rim. (Reproduced courtesy of Luc Viatour.)

Giora Shaviv (Bahcall 1963, 1964; Bahcall & Shaviv 1968) and detected by Ray Davis and collaborators (Davis et al. 1968) in an experiment that took place 1.5 km underground in the Homestake Gold Mine in Lead, South Dakota. The detector was comprised of a 380 cubic meter (100,000 gallon) tank of perchloroethylene, a common dry-cleaning fluid. In addition to ground- and underground observatories, several dozen spacecraft, starting in 1959 with NASA's Pioneer 5, many designed specifically to study the Sun, have given us an ever increasing understanding of our primary source of energy; they have also raised many questions that remain to be answered. The observatories have become increasingly more sensitive and the cadence of observations have also significantly increased (Figure 5.3).

Figures 5.4 and 5.5 illustrate the wealth of multi-wavelength information from the Sun as monitored by NASA's Solar Dynamics Observatory (SDO), which was launched in 2010. SDO has three instruments: the Helioseismic and Magnetic Imager (HMI), the Atmospheric Imaging Assembly (AIA), and the Extreme Ultraviolet Variability Experiment (EVE). The HMI, developed by the Stanford Solar Group provides maps of solar surface velocity (dopplergrams), photographs of the solar photosphere at 6173 angstroms with 1″ resolution, and maps of the photospheric magnetic field (magnetograms). The AIA, developed in collaboration with Lockheed Martin, observes the Corona and transitional regions (up to half a solar radius above the limb) in multiple wavelengths ranging from optical to X-ray, each wavelength detects a different temperature and hence probes different depths of the Sun with 1.5″ resolution; with a 10 second cadence, one second for each of 10 wavelengths. The EVE, developed in collaboration with the University of Colorado includes both normal and grazing-incidence grating spectrographs with unprecedented spectral resolution.

The Sun is usually a faint source of gamma-rays caused by cosmic rays hitting gas or lower energy photons in the Sun's atmosphere. But due to its proximity, our Sun can dominate the gamma-ray sky (Figure 5.6). These flares are caused by coronal mass ejections that result in charged particles traveling close to the speed of light:

Figure 5.3. This image illustrates the resolution capabilities of the Solar Dynamics Observatory (SDO; launched in 2010), the Solar Terrestrial Relations Observatory (STEREO; launched in 2006), and the Solar and Heliospheric Observatory (SOHO; launched in 1995). SDO's AIA instrument (right image) has twice the image resolution of STEREO (middle image) and four times greater imaging resolution than SOHO (left image). The image cadence also varies. SDO takes 1 image every second. At best STEREO takes 1 image every 3 minutes and SOHO takes 1 image every 12 minutes. (Reproduced courtesy of Solar Dynamics Observatory, NASA.)

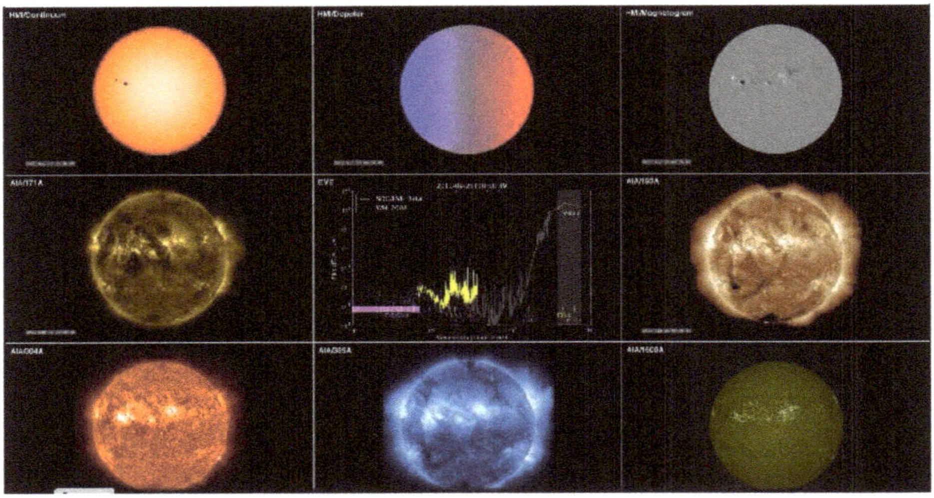

Figure 5.4. An animation covering 17 h of SDO observations. Each panel covers the same time interval. The upper three panels show a visible light image of the surface, a dopplergram, and a magnetogram as observed by HDI. The center panel shows the extreme ultraviolet spectrum using EVE, and the rest of the panels show five of the 10 wavelengths covered by the AIA. (Reproduced courtesy of Solar Dynamics Observatory, NASA.) Animation available at http://doi.org/10.1088/2514-3433/ac81b3.

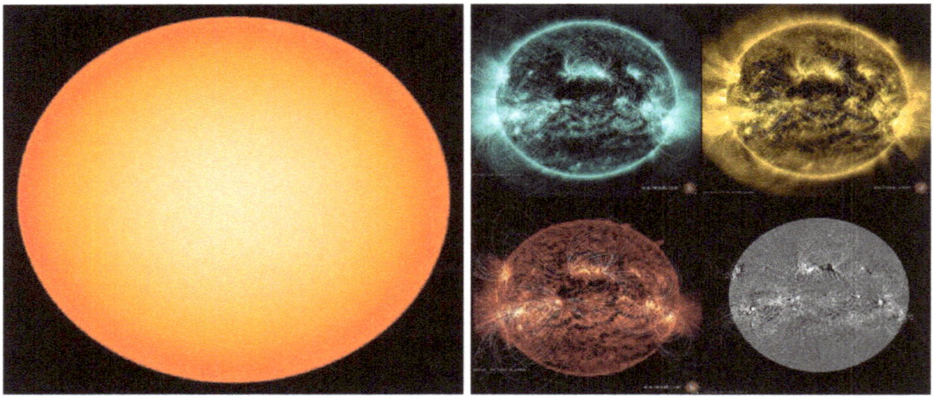

Figure 5.5. Left (a): A movie showing the rotating Sun for one week of a two week interval where it had almost no sunspots (first week of February 2018). It is hard to tell the Sun is rotating since there are very few features. This spotless period is a prelude to a period of solar minimum, when the Sun's activity will be at the low end of its 11-year cycle. (Reproduced courtesy of Solar Dynamics Observatory, NASA.) Animation available at http://doi.org/10.1088/2514-3433/ac81b3. **Right (b):** A representation of the Sun's magnetic field as measured with the HMI (lower right image) is generated for every AIA image. The HMI magnetogram illustrates how field lines emerge, extend high above the Sun and connect magnetic poles (black and white represent opposite polarity). (Reproduced courtesy of Solar Dynamics Observatory, NASA.)

some of these particles flow back onto the Solar surface along magnetic field lines and generate a flare of gamma-rays, temporarily making the Sun the brightest object in the gamma-ray sky.

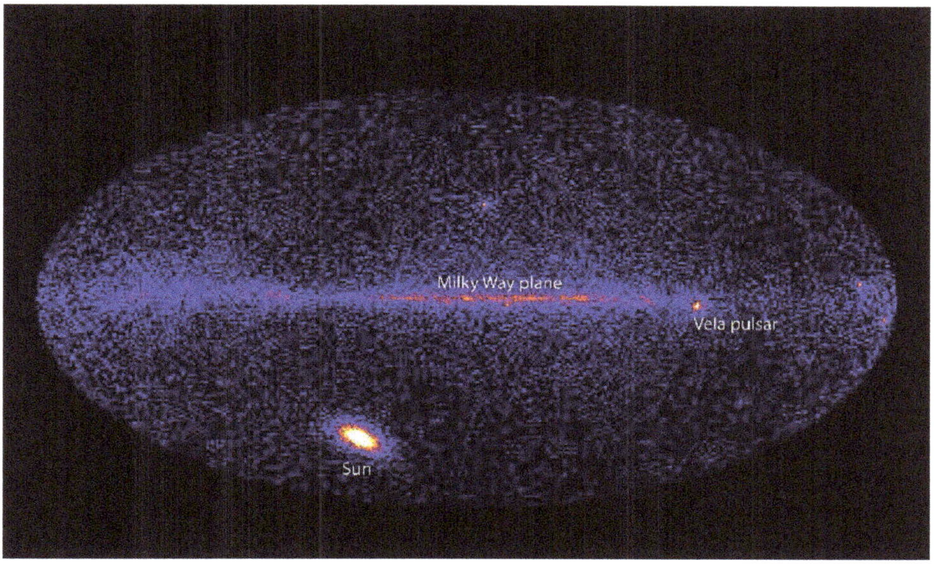

Figure 5.6. This image and animation shows a solar gamma-ray flare observed with the LAT on Fermi on 2012 March 12. It is the highest energy emission (4GeV) ever association with a Solar eruption. The flux was 1000 times greater than the Sun's steady output. (Reproduced courtesy of NASA, DOE, International Fermi LAT collaboration.) Animation available at http://doi.org/10.1088/2514-3433/ac81b3.

5.3 X-ray Binaries: The Most Luminous Objects in Our Galaxy

The first non-solar X-ray source discovered was an X-ray binary (Giacconi et al. 1962, 1964), although it's binary nature was not apparent at the time. Ricardo Giacconi, of American Science & Engineering in Cambridge, MA, spent nearly fourteen years after the detection of X-rays from the Sun in 1948 (Friedman et al. 1951) before he could get support to launch a rocket that would look for X-rays from something other than the Sun. Not surprising, since the Sun is a million times less bright in X-rays than in the optical and to detect such emission from even the closest star to us was simply not possible with the detector technology available in the 1950s. Giacconi obtained support to look for X-rays from the Sun as reflected by the Moon. During the 300 s the rocket was above the atmosphere, Giacconi took a quick sweep of the whole sky as well as pointing at the Moon (Figure 5.7(a)). He found a source that was a million times as powerful in X-rays as the Sun. That was the X-ray binary, Scorpius X-1/V818 Scorpi.[1] Giacconi et al. also found excess

[1] The early naming convention for X-ray sources was to assign a number and add the name of the constellation that defined the direction toward which the source was detected. Scorpius X-1 is the first X-ray source found in the direction of Scorpius. Ultimately, the number of X-ray sources increased as did the accuracy to which their position was determined. The current convention is to assign a label that identifies the satellite that first detected the source, and then the source position in Galactic coordinates. Scorpius X-1 is also known as 4U1615-17 where the 4U stands for the 4th catalog of sources from the X-ray satellite Uhuru launched by NASA in 1970. V818 Scorpi, the variable star that is paired with the neutron star to form the binary system had been known to optical astronomers for years.

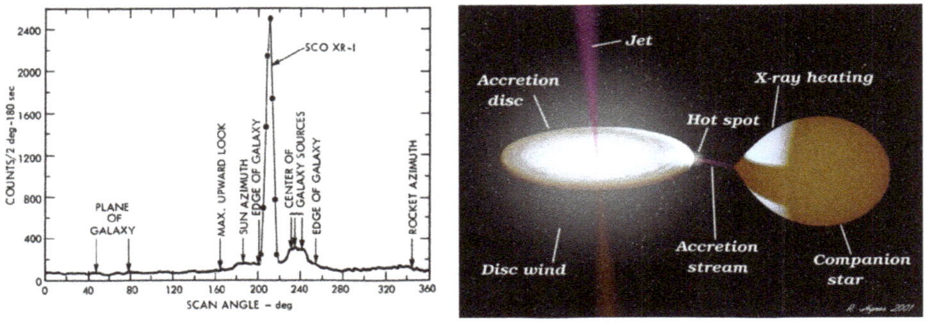

Figure 5.7. Left (a): Five minutes of data from a rocket-born X-ray detector flown by Giacconi in October 1967. This shows the counting rate of the detector as it scanned a great circle containing the source Sco X-1 and a cluster of sources in the direction of the galactic center. The detector field of view was 5 × 30 degrees. The Sun was below the horizon. (Reproduced courtesy of Fred Seward; Charles & Seward 2010.) **Right (b):** A schematic showing parts of an X-ray binary. Note the elongated shape of the companion star due to the rotational forces. The cusp is the inner Lagrangian or L1 point. (Reproduced courtesy of NASA/R. Hynes.)

X-ray emission from the Galactic center and background X-rays from all directions in the sky; both these phenomenon are still actively studied today.

XRBs are interacting binary systems comprising a "normal" donor star transferring matter on to a compact object such as a stellar-mass black hole or a neutron star (Figure 5.7(b)). The compact objects themselves are of astronomical significance as the endpoints of stellar evolution. The neutron stars and black holes that are key ingredients in XRBs form when normal stars die in a supernova explosion. Supernova cores that are greater than 1.4 solar masses form neutron stars, cores with greater than 3 solar masses form black holes. If left to themselves, neutron stars and black holes are essentially static over the lifetime of the universe. However, if you put them in a binary system with a companion that is willing to donate a bit of mass you can resurrect them in spectacular ways.

Neutron stars were defined in Chapter 4 (New Characters on the Cosmic Stage). Black holes are systems that are even denser than neutron stars, so dense that not even light can escape from them. Isaac Newton (1642–1727) noted that if you threw something fast enough from the Earth's surface it will go into orbit or even escape. The escape velocity of a massive body is the speed an object has to have before it can escape the massive body's pull. If a body is so massive that its escape velocity is greater than the speed of light it means even light cannot escape it. That means it cannot be seen: hence the term black hole.[2] The solution to Einstein's theory of general relativity developed by Karl Schwarzschild predicted such objects (Schwarzschild 1916) although Einstein himself was loathe to accept the idea. By the mid-1960s several theoretical studies had reached the same conclusion: that the

[2] The term "black hole" was adopted in 1967 by John Archibald Wheeler, following a suggestion from a member of the audience at a lecture he gave at the Goddard Institute for Space Studies in New York City (Wheeler 1998).

existence of black holes is unavoidable if we accept Einstein's general relativity (e.g., Thorne 1965). Just as a surprising number of cosmic objects seem to fall into two distinct categories, black holes have two extremes of size: stellar-mass black holes have masses of 3–20 times the mass of the Sun, whereas the supermassive black holes in the centers of galaxies and quasars range from millions to billion times the mass of the Sun. Although many candidates exist for medium sized black holes (e.g., Lael et al. 2022), there certainly does not appear to exist a full spectrum from 3 solar masses to a billion solar masses.

So what makes XRBs so much brighter in X-rays than our Sun? A star, such as our Sun, exerts a certain gravitational pull on the objects around it. The gravitational pull decreases as we go outward from the star. If there are two stars orbiting each other, there is one point between the two which feels equal pull from each, the inner Lagrangian point (Figure 5.8). If one of the stars expands to reach this point then material from it can start spilling through this point onto the other star. There are several times in every star's life when it passes through a giant stage

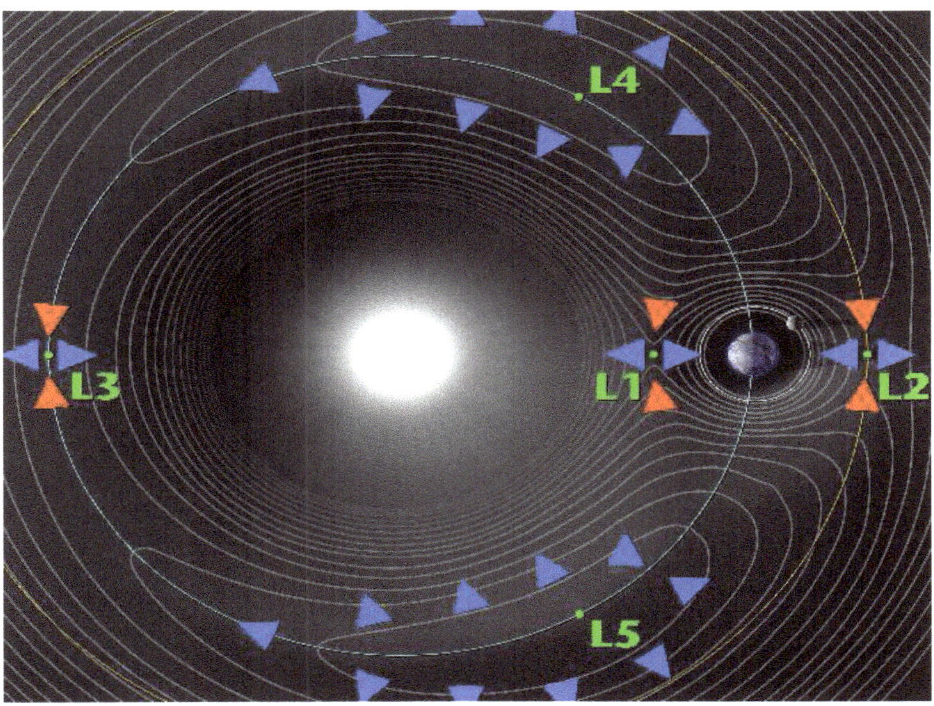

Figure 5.8. A schematic showing the effective potential for two bodies orbiting each other. Illustrated with our Sun and Earth. In a X-ray binary system the Sun would represent the companion "normal" star and Earth the compact body. The five points of equilibrium are called the Lagrangian points (although the first three were defined by Leonhard Euler). Material at the inner Lagrangian point (L1) is in equilibrium between the two masses. L1, L2, and L3 all have the same orbital period. L4 and L5 are stable and points where material will gather; the Trojan satellites of Jupiter are asteroids captured by the L4 and L5 points of the Sun-Jupiter Lagrangian. (Reproduced courtesy of NASA.)

(see Chapter 2: From Dust...The Birth and Death of Stars). When the normal star expands to the inner Lagrangian point, L_1, then the matter from the giant star is pulled at increasing velocities toward the compact object.

Consider a penny dropped from the 100th floor of the Empire State Building; it has considerably more impact than one dropped from the 2nd floor. If we drop a penny from just outside the Earth's atmosphere the potential energy it gains in the fall is that used by a 100W lightbulb burning for 1 h. If we throw a penny from the same distance above a neutron star the gravitational potential energy generated is 300 million times as much (the output of a large electric generating plant for half a day). In this case the matter is dropping substantially farther and onto an object that has a density 100 trillion times that of Earth.

In compact XRBs, gas from the atmosphere of the less dense companion is pulled at increasing velocities toward the super-dense compact star. Because the stars are orbiting each other, conservation of angular momentum means the gas cannot fall directly onto the compact star, but forms an encircling disk of material. As matter spirals though this accretion disk, it is accelerated and heated (viscous heating), causing it to radiate at ever shorter wavelengths until near the compact object it emits X-rays and γ-rays. Thus while XRBs were first discovered because they *are* bright in X-rays, they emit energy over a broad range of the electromagnetic spectrum.

Several features of XRBs are shown in Figure 5.7(b). A huge corona of hot gas above the heated disk. Sometimes particles are blown off the disk in a way similar to the Solar wind. Some of the X-rays generated heat up the disk and the normal star where it is not shaded by the disk. The accretion disk is not always enough to get rid of the angular momentum of rotation, so often huge jets of particles are thrown out perpendicular to the disk. XRBs are so bright that using the exquisite spatial resolution of the Chandra X-ray satellite they can be individually distinguished in other galaxies (Figure 5.9).

The first XRB system for which jet emission was suggested was SS433.[3] Several investigators proposed that the simultaneous red and blueshifts observed in the system were caused by jets moving in opposite directions (Milgrom 1979a,1979b; Martin & Rees 1979; Fabian & Rees 1979). They suggested the observed periodic velocity modulation was due to rotation of the jet axis. As the jets come toward the Earth they are blueshifted and as the jets move away from the Earth they are redshifted. Abell & Morgan (1979) using observations of Doppler shifts, found them consistent with a 164-day cycle of precession by the jets (Figure 5.10). This is seen in both X-ray and optical emission. Figure 5.10(a) shows a spectrum showing line shifts consistent with seeing the approaching (blueshifted) and receding (redshifted) jet at the same time (Eikenberry et al. 2001).

Radio observations of the transient X-ray source GRS 1915+105 following an unusually bright radio burst showed a pair of "blobs" emerging in opposite directions from a compact core (Mirabel & Rodriguez 1994; Figure 5.11(a)) over a period of weeks. Observing it at frequent intervals with the VLA, Luis Rodriguez

[3] Star number 433 in the Sanduleak and Stephenson 1977 catalog. There is no single, standard naming system for stars.

Figure 5.9. The upper-left box is an optical image of the Andromeda galaxy (M31) taken with NSF's 0.9 m telescope at Kitt Peak. The lower-left inset box is a Chandra X-ray image of the nucleus. The bright spots are mainly XRBs, a few are Ultra Luminous X-ray sources (these lie between the XRBs and AGN in luminosity), and a few are likely to be background quasars.. (Reproduced courtesy of Optical: NOAO/AURA/NSF/T.A. Rector & B.A. Wolpa; X-ray: NASA/UMass/Z. Li & Q.D. Wang.)

of the Institute of Astronomy at the National Autonomous University in Mexico City and Felix Mirabel of the Centre de Etudes at Saclay, France discovered the "blobs" were traveling at speeds apparently greater than that of light—a phenomenon called "superluminal motion" (see Section 3.1). Correcting for relativistic effects, Mirabel & Rodriguez calculated that the material in the jets of GRS 1915 +105 was actually moving at 92 percent of the speed of light. Although this makes the speed physically possible, it is still highly relativistic and requires enormous amounts of energy!

Superluminal jets had been observed in several quasars (see Section 4.2) but GRS 1915+105 was the first object found in our own Galaxy that displayed such "superluminal" speed. The similarity to quasars led Mirabel to term the objects "microquasars" (Mirabel 1993). It was an exciting result and images of the moving blobs made the cover of the journal Nature on 1994 September 1 (Mirabel & Rodriguez 1994).

In July 1994, the Compton Gamma Ray Observatory discovered an X-ray source in outburst (GRO J1655-40; Zhang et al. 1994; Figure 5.11(b)). Within two weeks of the X-ray burst a radio outburst was detected. Six days preceding the outburst an optical precursor had been observed by Orosz et al. (1997). Nine days following the radio

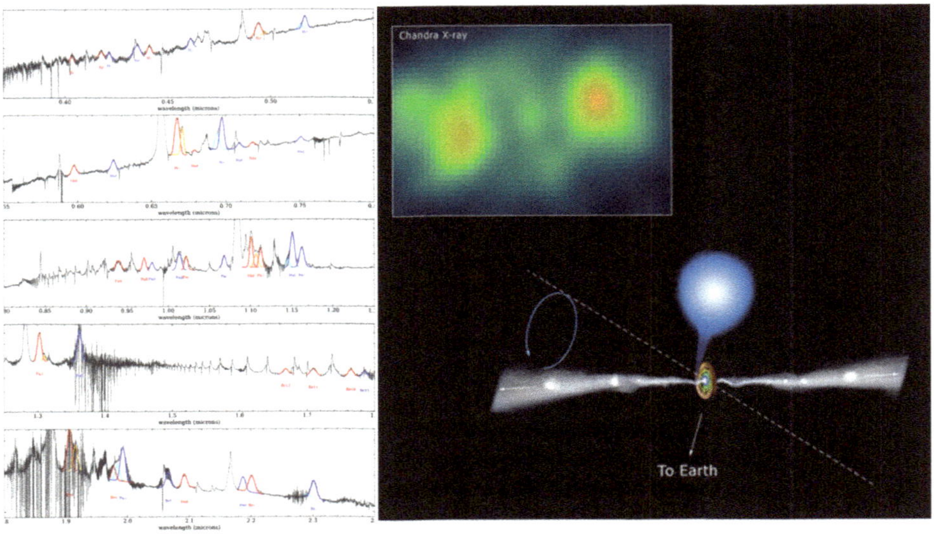

Figure 5.10. Left (a): A spectrum of SS 433 from the X-shooter Spectral Library. H I and He I emission lines from the eastern and western optical jets are shown in blue and red, respectively. Older jet ejections (earlier blobs; cyan and orange), are only present for the strongest line transitions such as Hα (second panel from top), are also shown. (Reproduced courtesy of Eikenberry et al. 2001.) **Right (b):** Upper-left inset showing lobes of X-ray emission from SS433 and lower panel a schematic model of the system. (Individual images courtesy of NASA).

outburst Tingay et al. (1994) undertook high resolution radio observations with the VLBI, and found two components moving away from each other at superluminal speeds. Hjellming and Rupen (1995) using the VLBA discovered a complex set of motions and variations suggesting the outgoing material in the jets follows a "cork-screw" pattern, rotating around the central axis of the jets about every three days. Dynamical evidence for the presence of a black hole in GRO J1655-40 was provided by Charles Bailyn and collaborators (Bailyn et al. 1995). Using the 4 m optical telescope at Cerro Tololo International Observatory (CTIO) in Chile, they found an eclipsing spectroscopic binary with a period of 2.62 days. Eclipsing meant the two objects passed in front of each other while orbiting and so were viewed edge-on. Spectroscopic binary meant lines from both objects could be seen to shift with the velocity of the objects movement. The mass of the larger object was found to be greater than 3 Solar masses, confirming it as a black hole. The orbital period was very close to the period of the corkscrew motion in the jets seen by Hjellming and Rupen. In 1999, Robert Hjellming of NRAO, took VLA observations of an object within 24 h of a reported X-ray outburst and found the object to be a microquasar (Hjellming et al. 2000).

5.3.1 Superluminal Motion

When matter approaches the speed of light relativistic effects become important. When objects are far away from us there is a time delay between the time we observe an event and the time that the event actually occurred. If the object is aimed at an

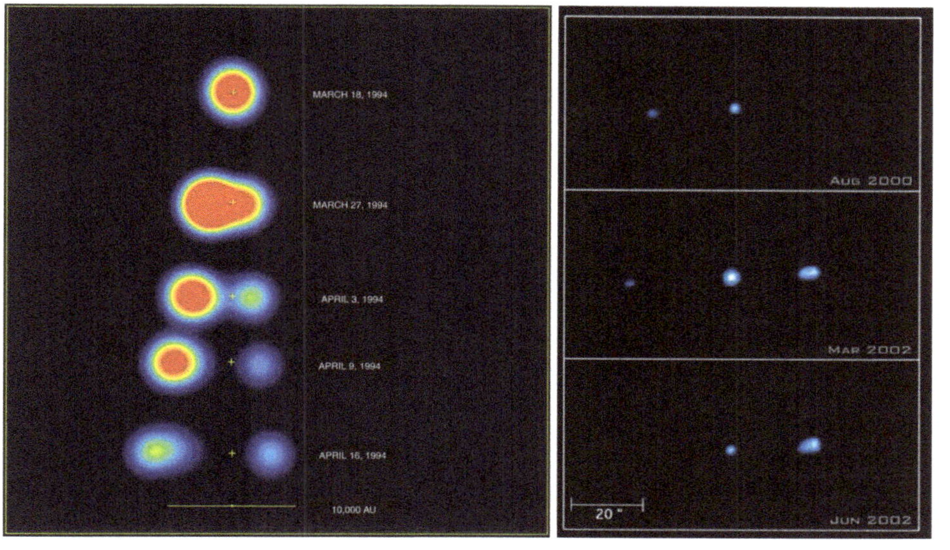

Figure 5.11. Left (a): It shows a pair of objects ejected from GRS 1915+105 that were moving apart apparently faster than the speed of light (superluminal speed). This was the first time that superluminal motion had been detected in our own Galaxy. The material ejected from the central object is moving away from the source over a period of weeks. (Reproduced courtesy of NRAO/AUI/NASA/Mirabel.) **Right (b):** An outburst of X-rays from the source, XTE J1550-564, was detected by NASA's Rossi X-ray Timing Explorer (RXTE) in 1998. This series of images from Chandra shows first one jet (left), then another opposing jet (right) of high-energy particles moving away from the black hole at about half the speed of light. Four years after the outburst, the jets had moved more than 3 light-years apart with the left jet slowing down and disappearing. (Reproduced courtesy of NASA/Chandra; Illustration credit: M. Weiss (CXC).)

angle to us then for the component of velocity directed toward us the time delay decreases. Figure 5.12 shows an example where an apparent speed of 1 and 1/2 times that of light (1.5c) is in reality only 5/6th the speed of light or 0.84c. It is the velocity component directed toward us that is the most effected and hence objects coming directly toward us can more easily show superluminal motion.

5.4 Quasars and Active Galactic Nuclei: The Most Luminous Objects in the Universe

Even before the detection of highly luminous X-ray point sources, astronomers in the late 1940s discovered radio emission up to a million times more powerful than that from our own Galaxy coming from galaxies billions of light-years away. In 1948 a variable source of radio emission was detected in the constellation Cygnus (Bolton & Stanley 1948). In 1952, the source was found to be at the center of a faint elliptical galaxy (Baade & Minkowski 1954). Even using the largest telescope available at the time, the 200-inch reflector at Mount Palomar Observatory in

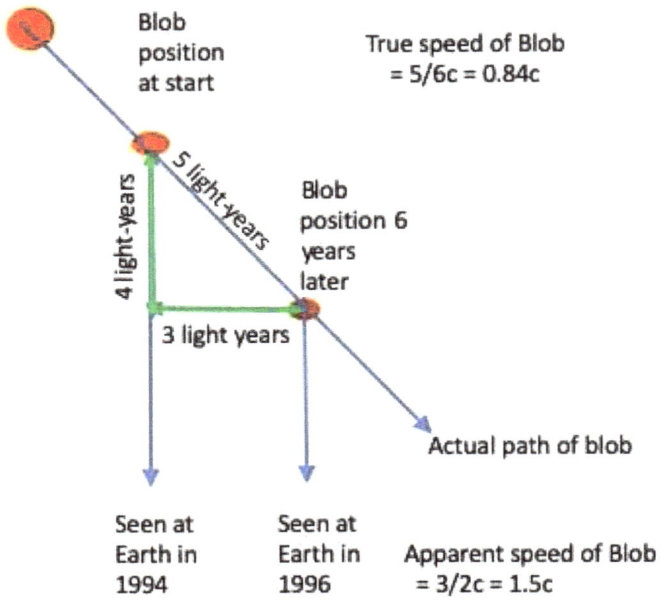

Figure 5.12. A schematic showing a "blob" ejected from a central source. The blob is moving at an angle to the observer. In this example the blob has traveled 5 light-years in 6 years. Hence its speed is a fraction of the speed of light: 0.84c. The blob (seen when it lights some intergalactic material) is detected on Earth in 1994 and in 1996. It appears to be moving 3 light-years in two, giving an apparent speed of 1.5c. However the 1996 position is 4 light-years closer than the 1994 position.

California, the Galaxy, now known as Cygnus A,[4] was barely visible. Cygnus A, and many galaxies first found in radio wavelengths were later found to have two extended radio lobes 10^5 to 10^6 light-years wide symmetrically placed about them (Jennisen & Das Gupta 1953; Figure 5.13). The giant lobes are formed when these jets are slowed down by the tenuous gas that exists between galaxies.

In the early 1960s very strong radio emission was measured from what appeared in the optical as faint blue star-like objects. The first optical identification was possible by a fortuitous occultation by the Moon (Hazard et al. 1963). The radio source, 3C273, blinked out just like a star, but the "star" emitted a tremendous amount of radio emission and had just a little jet like protrusion (Figure 5.14(a)). Martin Schmidt used the Palomar 200-inch to measure a redshift of 0.16 to 3C273 (Schmidt 1963). The spectrum he observed was unlike anything seen in a normal galaxy: it had huge hydrogen emission lines very unlike the blackbody spectra of the stars that compose normal galaxies. A substantial group of quasars with large redshifts, strong optical emission lines, and strong radio emission, but a star-like

[4] Early on radio sources were labeled by alphabetic letters and the constellation the source was in, hence Cygnus A. The naming convention for galaxies and nebulae is usually the name of a catalog and a number. NGC 4258 is object number 4258 in the New General Catalogue of Nebulae and Clusters of Stars. 3C273 is the 273rd object in the 3rd Cambridge Catalog of radio sources. M87 is the 87th object in the Messier catalog of nebulae. M87 is also NGC 4688 or Virgo A.

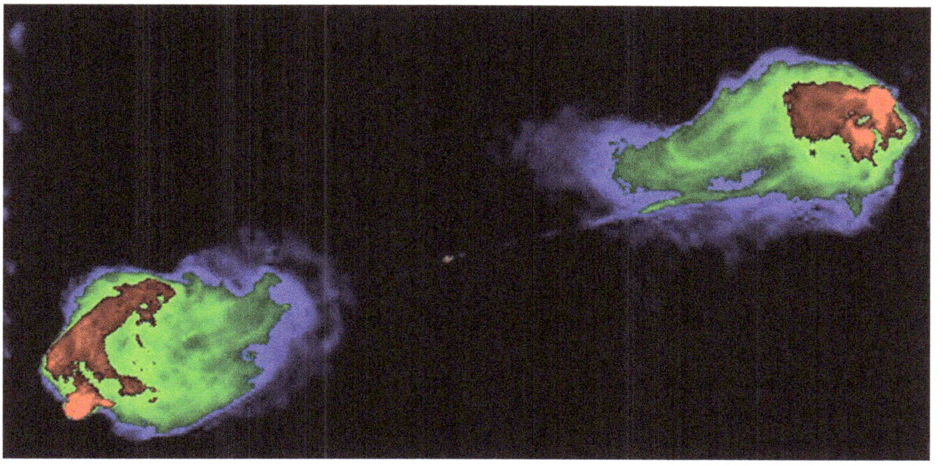

Figure 5.13. False color image of the radio jet and lobes in the hyperluminous radio galaxy Cygnus A. Red shows regions with the brightest radio emission, while blue shows regions of fainter emission. The giant elliptical galaxy, at a distance of 200 Mpc is not visible. At this distance, the lobes are separated by more than 100 kpc, and they have a radio luminosity of 10^{45} erg s^{-1}, a million times greater than the radio luminosity of normal galaxies. It is the closest powerful radio galaxy and one of the brightest radio sources in the sky. (Reproduced courtesy Mhardcastle.)

Figure 5.14. Left (a): The first discovered quasar, 3C273, as imaged with the Wide Field and Planetary Camera 2 (WFPC2) on Hubble. The little jet like projection barely seen with the Palomar shows here as a highly collimated jet 2×10^{18} km long. 3C273 is located in a giant elliptical galaxy in the constellation of Virgo. (Reproduced courtesy of ESA/Hubble/NASA.) **Right (b):** This Hubble image of M87 is a composite of individual observations in visible and infrared light. Its most striking features are the blue jet near the center and the myriad of star-like globular clusters scattered throughout the image. The jet is a black-hole-powered stream of material that is being ejected from M87's core. As gaseous material from the center of the Galaxy accretes onto the black hole, the energy released produces a stream of subatomic particles that are accelerated to velocities near the speed of light. (Reproduced courtesy of NASA, ESA and the Hubble Heritage Team (STScI/AURA); Acknowledgment: P. Cote (Herzberg Institute of Astrophysics) and E. Baltz (Stanford University).)

appearance were found in optical wavelengths. The term "quasi-stellar radio sources" applied to these objects was quickly shortened to quasars[5] by Hong-Yee Chiu (Chiu 1964). Subsequently, an even larger number of systems with large redshifts, strong optical emission lines, a star-like appearance, but little or no radio emission were found. Most quasars are "radio quiet," i.e., they have very little radio emission, whereas radio-loud quasars have strong radio jets, but the term quasar is commonly used to include both radio-loud and radio-quiet groups. Radio quiet galaxies are also called Seyfert galaxies after the person who first classified them (Seyfert 1943) or simply quasi-stellar objects (QSOs).

The first high redshift quasar found was actually 3C48 with a redshift of 0.37. Jesse Greenstein (1909–2002) who found it, considered the possibility of 0.37c but he did not believe his result, since no galaxies had been found at that distance at that time. Greenstein's paper cited Schmidt (Greenstein & Matthews 1963). It was therefore Schmidt who had his picture published in the cover of TIME magazine on 1996 March 11 as "Exploring the edge of the Universe" (The inferred distance to 3c273 was only 700 Mpc). There are many galaxies that contain "star-like" centers that are highly luminous. Galaxies with such quasars at their centers became known as active galactic nuclei (AGN). Figure 5.22(a) shows the spectra of a normal galaxy, a typical quasar, two types of Seyfert galaxies, and several other types of AGN. A model that accounted for radio-loud quasars was first developed by Meg Urry (Urry & Padovani 1995) and has since been modified to include nearly all classes of AGN only a few of which are discussed here (Figure 5.22(b)).

We now know that many (perhaps all) quasars are small regions of intense activity within otherwise normal galaxies. What is responsible for all the energy that quasars are seen to be producing—sometimes hundreds of times the energy from entire normal galaxies? The presence of radio lobes up to 10^6 light-years in size indicate that the central source of energy has been active at least that long. Since it is unlikely that the entire mass will be converted to energy with 100% efficiency the quasar has to be at least 10^8 solar masses. Quasars appear to us as point sources. The fact that they can vary on the order of days means they must be compact. Coupled with the very high masses involved this translates to supermassive black holes (SMBHs). Well-collimated jets of relativistic particles that extend for several thousand light-years emerge from the very centers of many active nuclei and quasars, and black holes are suspected to be the source of their power. All such systems also have disks, which is consistent with matter spiraling onto the SMBH. It is logical to assume that these SMBHs produce power by accreting material just as XRBs do. However, SMBHs do not come with companions from which they can pull mass (Figure 5.23) hence the source of the material being accreted is still in question. It is possible that the fuel is just stars that get too close. It is also possible that the energy we detect is being extracted from the SMBH in ways that we do not yet understand.

[5] Quasars were first found in radio wavelengths. They were as luminous as an entire galaxy but in optical light appeared star-like and so were called "quasi-stellar radio sources." In 1964, in an article written for Physics Today, Hong-Yee Chiu, shortened the term to "quasar."

With AGN as with XRBs the presence of disks has for the most part been inferred by indirect methods. Evidence for the presence of an accretion disk in the center of the spiral galaxy NGC 4258 was first detected by observing masers (molecules which amplify microwave radio emissions in the same manner that a laser amplifies optical light). The microwave emission is not attenuated by the dust and gas that surround many galactic nuclei and make study with optical telescopes difficult. Masers observed in NGC 4258 showed evidence of a rapidly rotating disk (Watson & Wallin 1994), determined the size and shape of the disk (Greenhill et al. 1995; Herrnstein et al. 1996; Figure 5.15(a)), and established the presence of a black hole of 4×10^7 solar masses (Miyoshi et al. 1995). VLBI observations obtained two years later (Herrnstein et al. 1997) showed that the masers had moved 1 milliarcsecond on the sky. As noted by Jim Moran in a NRAO press release "equivalent to a motion of about one millimeter seen at a distance of 3000 kilometers." These observations showed the exact location of the object around which the masers rotated and combined with observations that traced the inner parts of the jets provided a measurement of the distance between the black hole and the jets. Most satisfying, the offset measured in NGC 4258 was consistent with a prediction by Blandford & Konigl (1979). The measurements also established a geometric distance to NGC 4258 precise enough to provide an independent check on the extra-galactic distance-scale (Herrnstein et al. 1999).

An odd feature of NGC 4258 is the presence of two sets of spiral arms, that appear nearly perpendicular to one another. One set is observed only in visible and infrared, the other set only in radio and X-rays (Ogle et al. 2014; Figure 5.16). Centaurus A is another galaxy in which there is prominent radio and X-ray emission

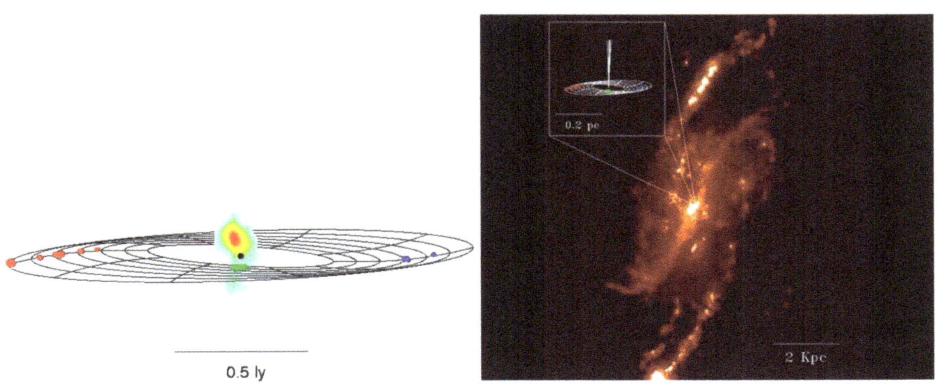

Figure 5.15. Left(a): A schematic showing the "central engine," of NGC 4258, a black hole surrounded by a very thin, slightly warped, molecular disk. The radio jets are shown in red, yellow, and turquoise (red is strongest radio emission and turquoise is weakest). The red, green, and blue spots represents masers. Red are moving away from us and blue are moving toward us circling around the black hole in the center. (Reproduced courtesy of James Herrnstein et. al., NRAO/AUI and the Harvard-Smithsonian Center for Astrophysics.) **Right (b):** An optical image of NGC 4258 showing the location of the disk. (Reproduced courtesy of NRAO/Gerald Cecil/Holland Ford.)

Figure 5.16. Upper panels: The X-ray (Chandra, blue) and radio (VLA, purple) profiles of the spiral galaxy NGC 4258 (also known as M106) are very different from the infrared (Spitzer, red) and optical (Hubble, yellow and navy blue) profiles. In visible and infrared two prominent spiral arms extend from the bright nucleus. Within the arms can be seen many bright young stars. But in radio and X-ray images, two additional spiral arms are seen that are nearly perpendicular with the optical/infrared ones. Radio observations also identified a pair of jets originating at the core and perpendicular to it (not shown). these are caused by gas that is being violently heated by shock waves. The jets heat the gas along their path forming an expanding cocoon, the cocoon generates shock waves that heats the gas in the disk to millions of degrees, causing it to radiate brightly in X-rays and radio. Lower panel: The four images combined show how the X-ray and radio emission lie perpendicular to the infrared and optical emission. (Reproduced courtesy of X-ray: NASA/CXC/Univ. of Maryland/A.S. Wilson et al.; Optical: Pal.Obs. DSS; IR: NASA/JPL-Caltech; VLA: NRAO/AUI/NSF.)

perpendicular to the optical disk (Figures 5.17 and 5.18). Towards the center of the Galaxy the arms appear to link-up with a north–south radio jet that presumably extends down in scale to the maser disk and ultimately the central engine. The anomalous arms are generally believed to be extensions of the relativistic jets originating in the central engine.

A possible "direct" view of an accretion disk from the elliptical galaxy NGC 6251 was obtained by HST by combining visible and ultraviolet light (Figure 5.19; Macchetto et al. 1997; Marconi et al. 1997). In optical light one sees a dark region associated with a disk of dust. In the ultraviolet one sees a bright emission from one side of the disk.

Figure 5.17. Centaurus A is another galaxy in which there is prominent radio and X-ray emission perpendicular to the optical disk. **Left (a):** A radio image of Centaurus A obtained with the Very Large Array. (Reproduced courtesy of NSF/VLA/Univ.Hertfordshire/M. Hardcastle.) **Right:** An optical image from the ESO's Wide-Field Imager (WFI) camera at the ESO/MPG 2.2 m telescope on La Silla, Chile. (Reproduced courtesy of ESO/WFI/M. Rejkuba et al.)

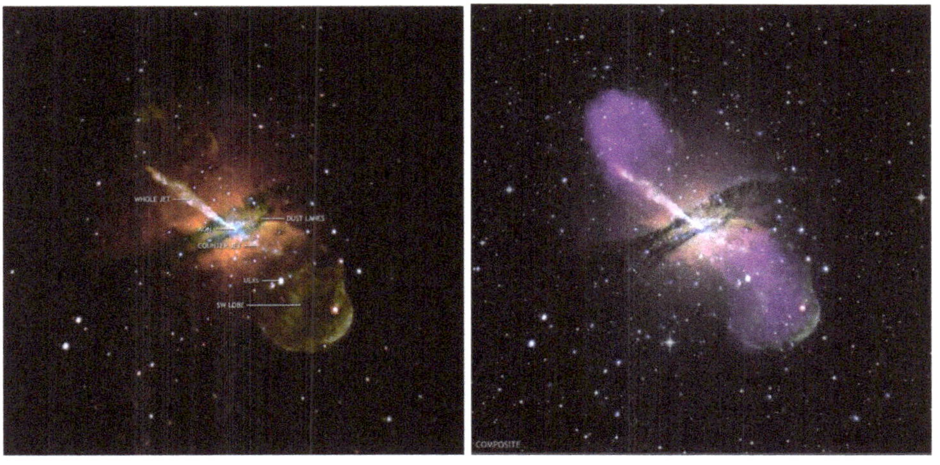

Figure 5.18. Left (a): An X-ray image of Centaurus A galaxy taken with the Chandra X-ray observatory. Red, green, and blue represent high, medium, and low X-ray energies. Features of the galaxy which are prominent in this X-ray image are marked: the opposing jets, the central engine, dust lanes, and ULXs (extra bright white spots). (Reproduced courtesy of NASA/CXC/CfA/R. Kraft et al.) **Right:** A composite of the radio, X-ray, and optical images. (Reproduced courtesy of NASA/CXC/CfA/R. Kraft et al.)

Water masers have been detected in several spiral galaxies, and as in NGC 4258 the masers are believed to lie within a disk of molecules orbiting the galaxy's central black hole. But recent radio observations of NGC 1052 (one of only two elliptical galaxies in which water masers have been found) show masers that lie along the "jet" of material being ejected by the central engine rather than in a disk. The shape of the

Figure 5.19. The left hand image shows a ground-based telescopic view of the core of NGC 6251, an elliptical galaxy that is 300 million light-years away in the direction of the constellation Ursa Minor. The inset box shows HST's field of view. The right hand side shows a composite image of the core of the galaxy that was constructed by combining a visible light image taken with Hubble's Wide Field Planetary Camera 2 (WFPC2), with a separate image taken in ultraviolet light with the Faint Object Camera (FOC). While the visible light image shows a dark dust disk, the ultraviolet image (color-coded blue) shows a bright feature along one side of the disk. Because HST sees ultraviolet light reflected from only one side of the disk, astronomers conclude the disk must be warped. The bright white spot at the image's center is light from the vicinity of the black hole which is illuminating the disk. (Reproduced courtesy of NASA/STScI.)

spectrum of NGC 1052's water masers is also different from that of the spirals; instead of having many narrow spiky features, they are broad and relatively smooth (Claussen et al. 1998).

5.4.1 A Neutrino from a Blazer; a Multimessenger Event

In the case where the jet from a quasar is aimed directly at us, it is called a blazer (Figures 5.22(b) and 5.23). The blazer, TKS 0506+056, has been identified as the source of an extremely high-energy neutrino (290 TeV) detected by the IceCube collaboration on 2017 September 22 (IceCube-170922A; IceCube Collaboration 2018; Figure 5.20). The Large Area Telescope on the Fermi gamma-ray satellite detected TKS 0506+056 in a flaring state (Tanaka 2017). IceCube's automatic notification system, alerted NASA's Neil Gehrels Swift Observatory. Swift undertook a 22 h survey in the direction specified by the IceCube within 3 h of the neutrino detection, and found nine X-ray sources, one of which was TKS 0506+056 (Keivani et al. 2017, 2018). Keivani et al. also used data from the Swift XRT and UVOT and NuStar and

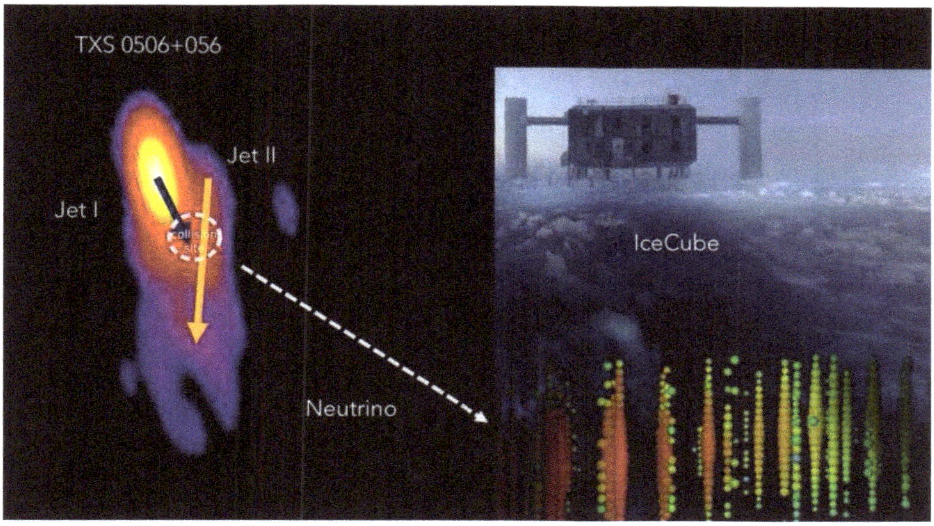

Figure 5.20. The left hand image shows TXS 0506+056 as observed with the VLBA. Britzen et al. (2019) suggest that IC-170922A originated in a precessing jet–jet interaction. The IceCube detection event is shown relative to the observatory. Each circle represents an excitation of a DOM. Bigger circles mean larger flux. Earlier times in red; later times in green. (Reproduced courtesy of the IceCube Collaboration/MOJAVE/S. Britzen/M. Zajacek.)

built a multimessenger model that could reproduce the spectral energy distribution over 10 orders of magnitude in photon energy as well as the neutrino detection. Searching back through its archive IceCube found several other neutrino detections that could be associated with TKS 0506+056. Meszaros et al. (2019) note that blazers may account for up to 30% of the IceCube neutrino background. This is one of the few cases where a specific astrophysical source has been associated with more than one messenger.

5.4.2 The Quasar/Microquasar Analogy

It is by now fairly well established that all galaxies have supermassive black holes at their centers (Figure 5.21; Neilsen et al. 2011). In a normal galaxy the black hole is relatively quiescent; the spectrum of a normal galaxy can be explained as a sum of the stars and dust in the galaxies (Figure 5.22(a)). The strong hydrogen emission lines, radio synchrotron emission and X-ray emission, that make active galaxies orders of magnitude more luminous are powered by material accreting on to the central black hole (Figure 5.22(a)). A model that accounted for radio-loud quasars was first developed by Meg Urry (Urry & Padovani 1995) and has since been modified to include nearly all classes of AGN only a few of which are discussed here (Figure 5.22(b)). However, the reason why some are radio-loud and produce massive jets whereas others are radio quiet is still under debate. The supermassive black hole at the center has a relatively thin accretion disk around it which is surrounded by a thick torus of dust. The various types of AGN are then due to the angle at which the

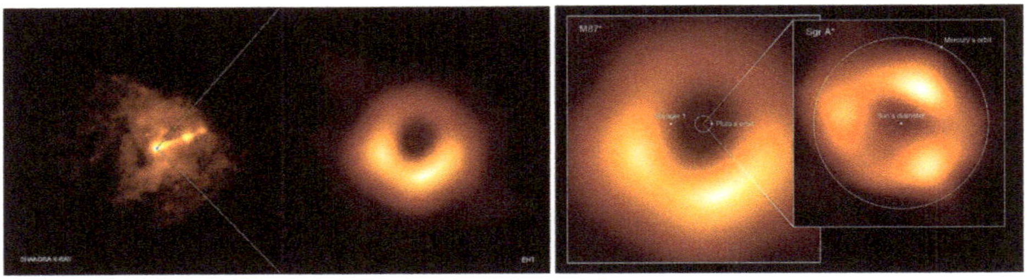

Figure 5.21. Left (a): A close-up of the inner region of M87 as observed with Chandra taken at the same time that the EHT obtained its image of its central supermassive black hole, M87* (between 3 and 7 $\times$ 10^9 Solar masses). (Reproduced courtesy of X-ray: NASA/CXC/Villanova University/J. Neilsen; Radio: Event Horizon Telescope.) **Right (b):** Image of the supermassive black hole (4 $\times$ 10^6 Solar masses) at the center of our Galaxy, SgrA*. (Reproduced courtesy of ESO/NASA.)

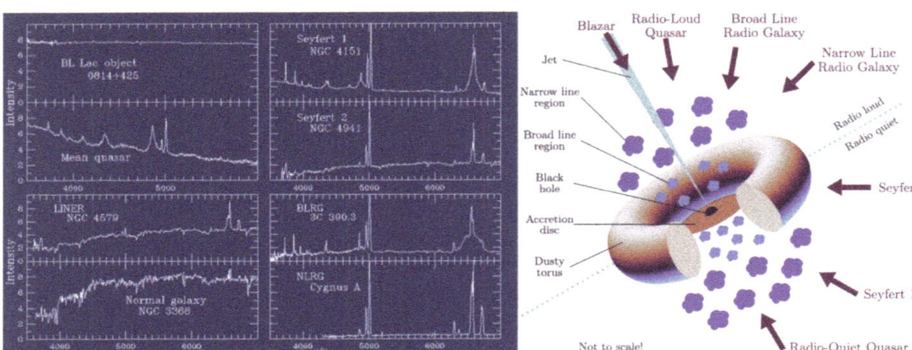

Figure 5.22. Left (a): A spectral array of spectra from various classes of AGN compared to a "normal" galaxy. This includes Seyferts which are radio-quiet systems, (Reproduced courtesy of William Keel.) **Right (b):** A unified model which suggests that all AGN a central thin disk around a black hole surrounded by a thick torus of dust and with clumpy material in a halo above the disk and torus. This explains the different categories of AGN as resulting from a different angle of view toward the central source. Only radio-loud sources produce jets but what causes some galaxies to be radio-loud and some to be radio-quiet is still under debate. (Reproduced courtesy of Emma Alexander.)

observer looks toward the central black hole. Broad-line radio galaxies (BLRGs in Figure 5.22) look directly in toward the SMBH, narrow-line radio galaxies are blocked by the torus. BLRGs are parallel to Seyfert 1s and NRLGs to Seyfert 2s.

The discovery of superluminal jets from XRBs means a remarkable analogy that can be drawn between XRBs and quasars (Figure 5.23). The physics of stellar-mass black holes in XRBs and the supermassive black holes in quasars scale both in size and time with the mass of the black hole (Rees 1998). Both display relativistic jets ejected in opposite directions from the central source. Though the neutron stars or black holes in microquasars are only a few times the mass of the Sun, the phenomena associated with them, such as the jets and superluminal position, are similar to those seen in active galaxies and quasars containing supermassive black holes. Because of

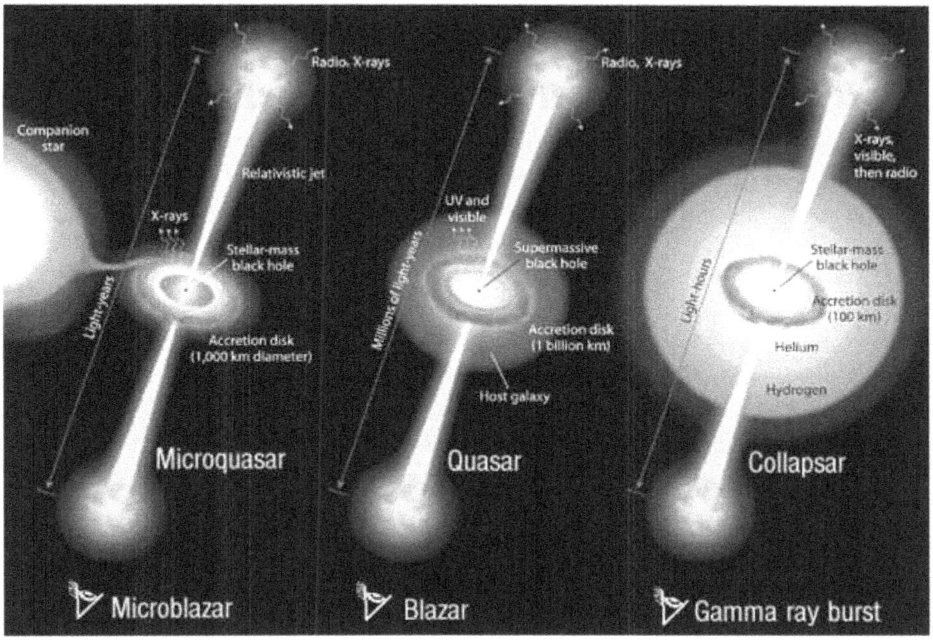

Figure 5.23. An illustration demonstrating the similarity of XRBs, quasars, and the collapsar model for long gamma-ray bursts. Blazars and microblazars are quasars and XRBs whose jets face us directly. A long gamma-ray burst appears when one looks directly into a jet that emerges from the collapse of a massive star (see Chapter 4). All three systems have disks. In XRBs the accreting material comes from a companion star, in collapsars the fuel is the massive star itself, the fuel for quasars are thought to be stars that get captured by the central black hole. (Reproduced courtesy of Mirabel 2003.)

their nearness and rapid variation (on the order of weeks) "microquasars" and "microblazers" offer an opportunity to study some aspects of relativistic ejections seen in the universe on shorter timescales than quasars and AGN that show variations over years. As such, microquasars provide much closer "laboratories" for study of these phenomena, which remain poorly understood.

5.5 Ultra High Energy Cosmic Rays (UHECRs): Microscopic Particles with Macroscopic Energies

Cosmic rays with energies above 10^{18} eV are called UHECR. The flux versus energy spectrum of cosmic rays (see Chapter 1, Figure 1.16) measured by the Pierre Auger observatory in Argentina and the Telescope Array UHECR Observatory in the US, is nearly featureless over 10 decades of energy. Small changes around 10^{15} eV and 10^{18} eV dubbed "knee" and "ankle" and a sharp drop-off about $10^{19.5}$ eV provide the only hints to their nature. The sharp cutoff could be due to the Greisen–Zatsepin–Kuz'min effect (GZK; Greisen 1966; Zatsepin & Kuz'min 1966): interaction of cosmic rays with the cosmic microwave background photons, which causes photodissociation and proton energy losses. An alternative explanation could be that the sources from which the cosmic rays originate cannot accelerate the protons to higher energies. Observations also suggested that the UHECRs are composed of heavier particles above about $10^{18.5}$ eV (Meszaros et al. 2019 and references therein). This conclusion is still under debate, and will only be solved when a statistically sounder sample of sources is obtained.

The lowest energy cosmic rays (below 20MeV) come from our Sun and ones up to the knee may be explained by supernova remnants where strong shocks are found (see Chapter 2). But supernova remnants cannot account for the highest energy particles. Both colliding galaxies and AGNs have been suggested as candidate sources. Since roughly 10 percent of all galaxies are AGN, and determination of the direction from which a given cosmic ray arrives is poorly resolved, an AGN can always be found within the very large regions associated with the arrival direction of a UHE cosmic ray. Although no correlations have been found with UHECR and known astronomical sources, the energy generation rates of UHECRs are comparable with PeV neutrinos and sub-TeV gamma-rays (Fang & Murase 2018). Fang and Murase suggest that supermassive black holes jets in clusters of galaxies could be the source of all three distributions (Figure 5.24). In the galaxy cluster

Figure 5.24. A composite image of galaxy cluster MS0 735.6+7421 showing jets emanating from the central galaxy. The optical image obtained with the Advanced Camera for Surveys on Hubble is depicted in white and shows the many dozens of galaxies that comprise the cluster. The intercluster gas, detected with the Chandra X-ray Observatory at temperatures of 50 million degrees, is depicted in blue. The giant cavities in the gas are filled by the relativistic radio jets observed by the VLA shown in red. (Reproduced courtesy of X-ray: NASA/CXC/Univ. Waterloo/B. McNamara; Optical: NASA/ESA/STScI/Univ. Waterloo/B. McNamara; Radio: NRAO/Ohio Univ./L. Birzan et al.)

MS0 735.6+7421, the lowest energy cosmic rays are confined within the radio lobes, medium-energy cosmic rays interact with the magnetized cluster environment to produce neutrinos and gamma-rays, and the highest energy cosmic rays escape the cluster.

References

Abell, G. O., & Margon, B. 1979, Natur, 279, 701

Baade, W., & Minkowski, R. 1954, ApJ, 119, 206

Bahcall, J. 1964, PhRvL, 12, 300

Bahcall, J., Fowler, W. A., Iben, I. Jr, & Sears, R. L. 1963, ApJ, 137, 344

Bahcall, J. N., & Shaviv, G. 1968, ApJ, 153, 113

Bailyn, C. D., Orosz, J. A., McClintock, J. E., & Remillard, R. 1995, Natur, 378, 157

Blandford, R. D., & Koenigl, A. 1979, ApJ, 232, 34

Bolton, J. G., & Stanley, G. J. 1948, Natur, 161, 312

Britzen, S., Fendt, C., Boettcher, M., et al. 2019, A&A, 630, 103

Charles, P. A., & Seward, F. D. 2010, Exploring the X-ray Universe (Cambridge: Cambridge Univ. Press)

Chiu, H-Y. 1964, PhT, 17, 21

Claussen, M. J., Diamond, P. J., Braatz, J. A., Wilson, A. S., & Henkel, C. 1998, ApJ, 500, L129

Davis, R., Harmer, D. S., & Hoffman, K. C. 1968, PhRvL, 20, 1205

Eikenberry, S. S., Cameron, P. B., Fierce, B. W., et al. 2001, ApJ, 561, 1027

Fabian, A. C., & Rees, M. J. 1979, MNRAS, 189, 13

Fang, K., & Murase, K. 2018, NatPh, 14, 396

Friedman, H., Lichtman, S. W., & Byram, E. T. 1951, PhRv, 83, 1025

Galileo, G. 1613, Istoria e dimostrazioni intorno alle macchie solari e loro accidenti (Rome: Giacomo Mascardi)

Giacconi, R., Gursky, H., Paolini, F. R., & Rossi, B. B. 1962, PhRvL, 9, 439

Giacconi, R., Gursky, H., Waters, J. R., Clark, G., & Rossi, B. 1964, Natur, 204, 981

Greenhill, L. G., Jiang, R. D., Moran, J. M., et al. 1995, ApJ, 440, 619

Greenstein, J. L., & Matthews, T. A. 1963, AJ, 68, 279

Greisen, K. 1966, PhRvL, 16, 748

Hale, G. E. 1908a, JGR, 13, 159

Hale, G. E. 1908b, PASP, 20, 287

Hale, G. E. 1908c, ApJ, 28, 315

Hazard, C., Mackey, M. B., & Shimmins, A. J. 1963, Natur, 197, 1037

Herrnstein, J. R., Greenhill, L. J., & Moran, J. 1996, ApJ, 468, L17

Herrnstein, J. R., Moran, J. M., Greenhill, L. J., et al. 1997, ApJ, 475, 17

Herrnstein, J. R., Moran, J. M., Greenhill, L. J., et al. 1999, Natur, 400, 539

Hjellming, R. M., & Rupen, M. P. 1995, Natur, 375, 464

Hjellming, R. M., Rupen, M. P., Hunstead, R. W., et al. 2000, ApJ, 544, 977

IceCube Collaboration, et al. 2018, Sci., 361, 1378

Jennison, R. C., & Das Gupta, M. K. 1953, Natur, 172, 996

Keivani, A., Evans, P. A., Kennea, J. A., et al. 2017, GRB Coordinates Network, Circular No. 21930

Keivani, A., Murase, K., Petropoulou, M., et al. 2018, ApJ, 864, 84

Lael, S., Wook, J-H., Donghoon, S., et al. 2022, AJ, 163, 15

Macchetto, F., Marconi, A., Axon, D. J., et al. 1997, ApJ, 489, 579
Marconi, A., Axon, D. J., Macchetto, F. D., et al. 1997, MNRAS, 289, L21
Martin, P. G., & Rees, M. J. 1979, MNRAS, 189, 19
Meszaros, P., Derek, B. F., Hanna, C., & Murase, K. 2019, NatRP, 1, 585
Milgrom, M. 1979a, A&A, 78, 3
Milgrom, M. 1979b, A&A, 78, 9
Mirabel, I. F. 1993, Annals of the New York Academy of Science, Vol. 688 (New York: Academy of Science), 581
Mirabel, I. F., & Rodriquez, L. F. 1994, Natur, 371, 46
Mirabel, I. F. 2003, NewAR, 47, 471
Miyoshi, M., Moran, J., Herrnstein, J., et al. 1995, Natur, 373, 127
Neilsen, J., Remillard, R. A., & Lee, J. C. 2011, ApJ, 737, 69
Ogle, P. M., Lanz, L., & Appleton, P. N. 2014, ApJ, 788, L33
Orosz, J. A., Remillard, R. A., Bailyn, C. D., & McClintock, J. 1997, ApJ, 478, L83
Rees, M. J. 1998, in Black Holes and Relativistic Stars, ed. R. M. Wald (Chicago, IL: Univ. Chicago Press), 79
Scheiner, C. 1630, Rosa Ursina (Bracciano: Paolo Giordano Orsini)
Schmidt, M. 1963, Natur, 197, 1040
Schwabe, S. H. 1843, AN, 21, 233
Schwarzschild, K. 1916, SPAW, 189
Seyfert, C. K. 1943, ApJ, 97, 28
Tanaka, Y. T., Buson, S., & Kocevski, D. 2017, ATel, 10791, 1
Tingay, S. J., Jauncey, D. L., Preston, R. A., et al. 1994, Natur, 374, 141
Thorne, K. S. 1965, Sci, 150, 167
Urry, C. M., & Padovani, P. 1995, PASP, 107, 803
Watson, W. D., & Wallin, B. K. 1994, ApJ, 432, L35
Wheeler, J. A., & Ford, K. 1998, Geons, Black Holes, & Quantum Foam: a life in physics (New York: W.W. Norton and Company)
Zatsepin, G. T., & Kuz'min, V. A. 1966, JETPL, 4, 78
Zhang, S. N., Wilson, C. A., Harmon, B. A., et al. 1994, IAUC, 6046, 1

Chapter 6

From Energy to Matter: Cosmology

Hubble image of gravitational lensing. Reproduced courtesy of NASA, ESA, K. Sharon (Tel Aviv University) and E. Ofek (Caltech).

6.1 Introduction

We know that we can get energy from matter, but where did the matter come from? In some sense, cosmology is our attempt to answer this question, to find why the universe is the way it is and how it got that way. What does it mean to understand the origin and evolution of our universe? What we can do is generate models of how the universe began and what causes it to evolve. The better a model is able to reproduce what we observe, the more predictions it makes that we can confirm, the more confidence we can have that we have a "correct" understanding of our observable universe. Certainly our understanding of the universe is changing with time: it is less than a century since we understood that there were galaxies outside ours; less than half a century since we accepted that there existed matter that we cannot see and our universe consisted of

doi:10.1088/2514-3433/ac81b3ch6

90% dark matter; now evidence points toward matter itself composing less than 30% of the universe with over 70% composed of something that we do not yet understand, something that has the opposite effect of gravity, and is referred to as "dark energy".

Multi-wavelength and multimessenger observatories provide a multitude of details that any model of the universe must be able to reproduce. From the formation of stars and galaxies, the expansion rate of galaxies, the clustering of galaxies causing inhomogeneities on the largest scales, dark matter, the cosmic microwave background (CMB), the minute fluctuations of the CMB, and the overwhelming preponderance of dark energy. It is a realm where not just multiple channels but multiple disciplines of science are utilized in the search for answers. Elementary particle physicists collaborate with astrophysicists and cosmologists.

6.2 The Expanding View of Our Universe

In the early 1600s, Galileo (1564–1640) had resolved many of the bright "nebulosities" in the Milky Way into individual stars. By 1750, Thomas Wright (1711–1786) had realized that the "fixed" stars are not just scattered randomly all around us but tended toward a plane (Figure 6.1(a)). By 1785, the Herschels (Sir William, 1738–1822, and his sister Caroline, 1750–1838) had actually counted stars in every direction. Although they measured stars extending farther in one direction than others they concluded the stars are distributed fairly uniformly with the Sun at the center (Figure 6.1(b)). Of course they did not know about dust and extinction so they did not realize that when we look toward the center of the Galaxy we see far fewer stars than are actually there. By 1915, Harlow Shapley (1885–1972) realized that we should be able to see Globular Clusters (GCs) to greater distances than individual stars since they are composed of thousands of stars packed together. He was also able to determine distances to GCs by using the large red variable stars they contained (Figure 6.2; Shapley 1917). He could see that the GCs appeared to be symmetric around a point rather far from the Sun and concluded, correctly, that we were closer to one edge of the total distribution. And he assumed that because the GCs are huge clusters of stars their gravitational distribution better defined the center of the Galaxy (Shapley 1918). Shapley concluded that our Sun was offset by a tremendous distance from the center of the Galaxy (Figure 6.3).

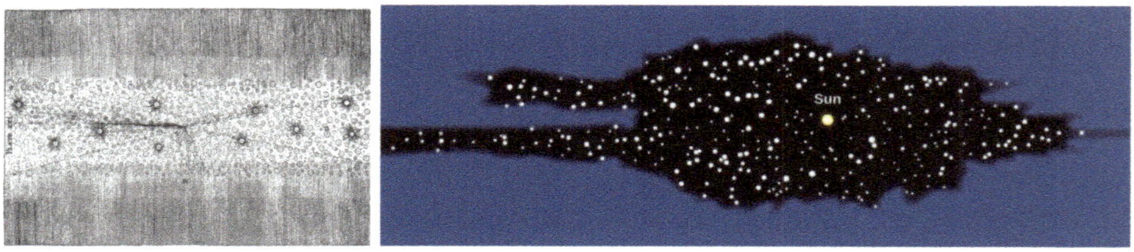

Figure 6.1. Left (a): Plate XXI from Thomas Wright's An Original Theory or New Hypothesis of the universe (1750) illustrating his concept of the universe: a uniform distribution of stars in a limited plane. (Reproduced courtesy of The Public Domain Review.) **Right (b):** The universe as visualized by Sir William Herschel in 1785 based on extensive observations by himself and his sister, Caroline Herschel. They concluded the stars are distributed fairly uniformly with our Sun roughly at the center, although they clearly observed stars in one direction extending farther for observations above and below the plane. (Reproduced courtesy of CNX/Mellema.)

Figure 6.2. This Hubble ACS image of NGC 362 shows the spherical structure of stars with sharply increasing concentration toward the center that is typical of globular clusters (GC). The large red variable stars used by Shapley to measure distances to the GCs can be seen distinctly. Stars in a GC are tightly bound by gravity in an area extending over 10e14 km (2 light-years) and containing 1000s of stars totaling masses up to a million Suns (10^{36} kg). Essentially all galaxies contain GCs; the Milky Way has about 150. (Reproduced courtesy of NASA/ESA/Hubble.)

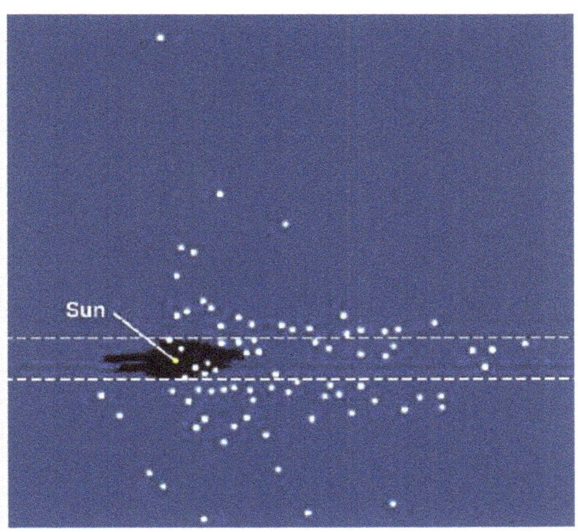

Figure 6.3. Left (a): Harlow Shapley. (Reproduced courtesy of CNX/Mellema.) **Right (b):** The distribution of Globular Clusters in our Galaxy as measured by Harlow Shapley in 1915. He measured the "universe" to significantly greater distances than the Herschels (their measured distribution from Figure 6.2(b) silhouetted in black) and concluded that our Sun was not at the center. (Reproduced courtesy of CNX/Mellema.)

Shapley actually thought our Galaxy was much larger than it was because he did not take into account the reddening of the GCs due to the Interstellar Medium. He thought the fainter GCs he observed were much farther away than they were in reality.

The "universe" at this time still consisted of just our Galaxy. But there had been observations of many spiral shaped "nebulae" within which individual stars had not been resolved in order to determine distances. Around this time, Henrietta Leavitt (1868–1921; Figure 6.4(a)), at the Harvard College Observatory, found a relationship of the extremely bright Cepheids that allowed for determining distances even farther than GCs (Leavitt 1908). Classical Cepheids are up to 30,000 times brighter than our Sun and highly variable. Leavitt found that the variability followed a distinct period for each individual object. At the dimmest part of a Cepheid's cycle, the gas in the outer layers of the star is heated by the star's radiation and is ionized so it cannot cool off by emitting light and the increase in temperature causes the star to expand. As it expands, it cools, and so becomes less ionized and therefore more transparent, allowing the radiation to escape. Then the expansion stops and reverses due to the star's gravitational attraction. The process then repeats. And it repeats in a remarkably consistent way. The more massive stars are bigger and therefore more luminous. And because they are bigger they take longer to expand and decrease so the period of expansion and reduction is increased. So the period is directly related to luminosity (Figure 6.4(b); Leavitt & Pickering 1912). Since periods are easier to measure than luminosity these stars became very useful for determining distances far greater than that possible with the red variable stars in GCs.

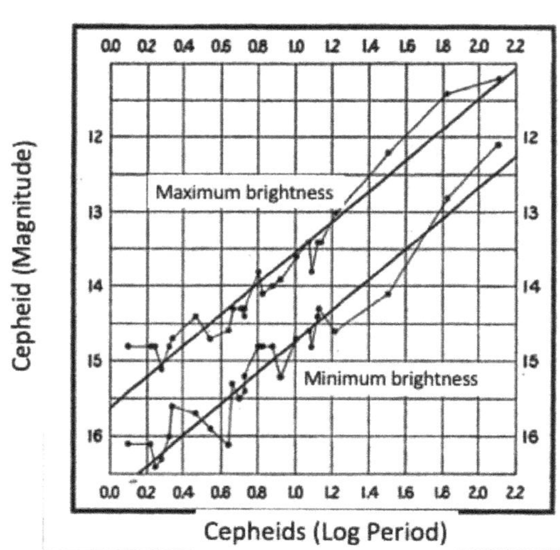

Figure 6.4. Left (a): Henrietta Leavitt (1868–1921) was the first to find the period–luminosity relationship of Cepheids that allowed measurement to distances far greater than that possible with red variables. (Reproduced courtesy of AIP.) **Right (b):** In her plot and words: "A straight line can be readily drawn among each of the two series of points corresponding to maxima and minima, thus showing that there is a simple relation between the brightness of the Cepheid variables and their periods." (Leavitt & Pickering 1912; Harvard College Circular 173.)

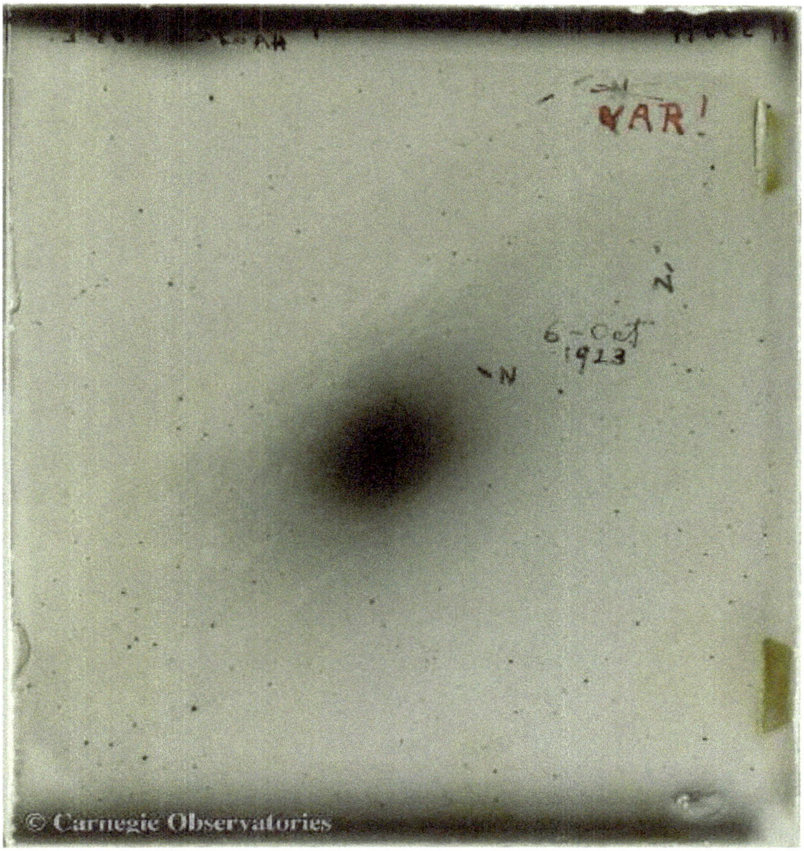

Figure 6.5. From nebula to galaxy: the observation that increased the size of the universe a millionfold. An image of the glass plate on which, in 1923, Hubble first identified (in bright red) a Cepheid variable in the then Andromeda Nebula. (Reproduced courtesy of Carnegie Institution for Science.)

In 1925 Edwin Hubble (1889–1953) identified a Cepheid variable in the Andromeda "nebula" and concluded that the nebula was at a distance significantly beyond the known extent of our own Galaxy (Figures 6.5 and 6.6; Hubble 1925). That one observation suddenly expanded our universe to something almost beyond imagination in the 1920s. Only forty years earlier in the 1880s we had first realized that our Sun was not unique and all stars were suns. Now not only was our Sun only one star among millions in the Milky Way: the Milky Way itself was only one galaxy among millions of galaxies in the universe (Figure 6.7).

6.3 The Cosmological Principle

The equations for Einstein's General Theory of Radiation (GR) were very complex and in order to form solutions for them Einstein (1879–1955) had to assume several simplifying assumptions: one that the universe was infinite, uniform, and static, the so-called "cosmological principle." At a time when there were very few physical

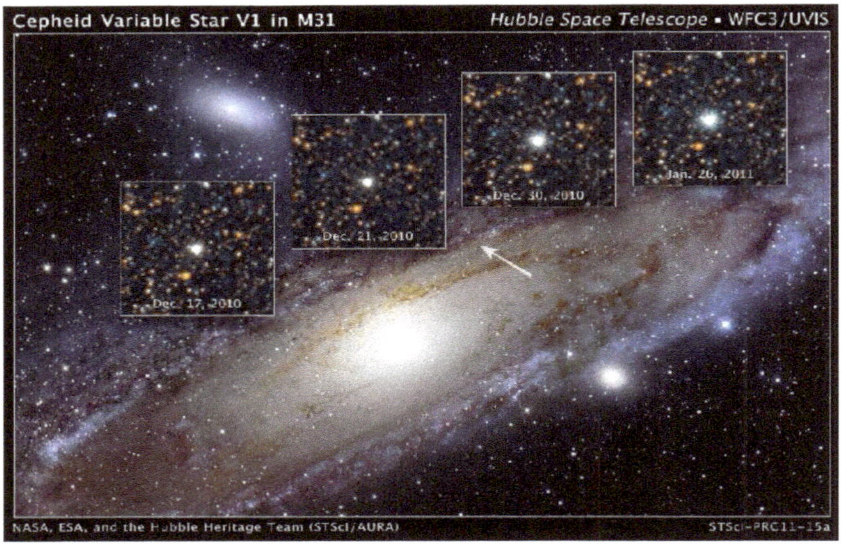

Figure 6.6. Insets show Hubble's Cepheid variable, V1, as observed with HST. At its dimmest on 2010 December 21; at its brightest on 2011 January 26. The Andromeda galaxy itself forms the background. (Reproduced courtesy of NASA, ESA, and the Hubble Heritage Team (STScI/AURA).)

Figure 6.7. Clusters of galaxies form the largest known gravitationally bound systems in the universe. Ranging in mass from 10^{14} to 10^{15} solar masses (up to 10^{45} kg) and in size up to 10^7 lys (10^{20} km). In this image of the galaxy cluster, Abell 1689, pictured by the Hubble Space Telescope, 100 million degree gas observed by the Chandra X-ray Observatory is overlaid in blue. The long arcs are caused by gravitational lensing. (Reproduced courtesy of NASA/CXC/MIT/E.-H. Peng et al.; NASA/STScI.)

observations of the universe, and when even the concept of galaxies had not been established, this was rather a big step. Especially since the cosmological principle was already very much in question through a paradox that many scientists were concerned about but that was highly popularized by Wilhelm Olbers (1758–1840) in the mid-19th century. The Olbers' paradox notes that the darkness of the night sky means that at least one of our assumptions must be wrong: either the universe is finite, or it is not uniform, or it is changing in ways that affect how light reaches us (Figure 6.8). If the universe is infinite, uniform, and static then all lines of sight should eventually hit a star. All stars are distant suns, and their combined brightness should be much more than the brightness of our Sun.[1] Distant stars are faint, but because of the inverse square law the reduction in brightness per unit area is exactly compensated by the increase in number of stars. It cannot be the intervening dust obscuring the stars, because the dust will heat up so it will either glow (like the dust we see in infrared in the center of our Milky Way) or it will evaporate like the dust from the early solar system.

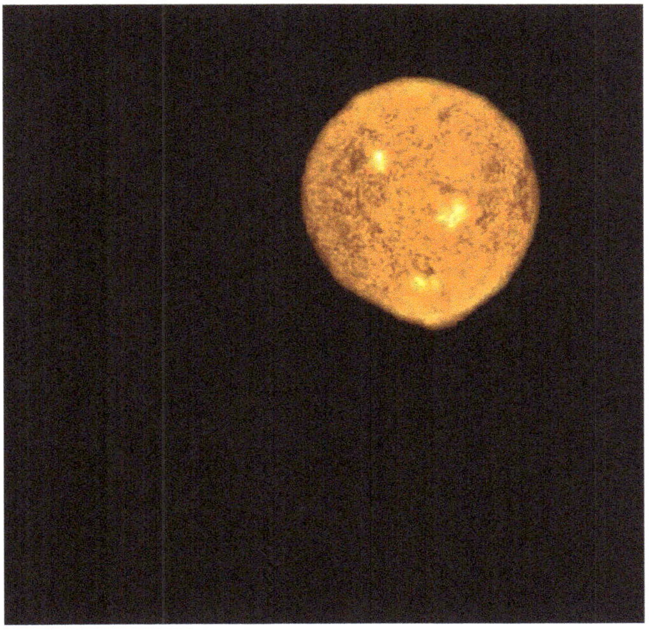

Figure 6.8. Animation illustrating Olber's Paradox: how the night sky should look if the universe is infinite, homogeneous, and static. (Reproduced courtesy of K. Marinas.) Animation available at http://doi.org/10.1088/2514-3433/ac81b3.

[1] The currently favored solution to Olbers' paradox involves the finite speed of light and the expansion of space itself. The finite speed of light means that light from all the stars in the universe has not had enough time to reach us. The expansion of space itself means that wavelengths increase with time and far away starlight is shifted away from the visible to the infrared.

Certainly, to the largest scales that we are able to observe we find that the universe is not uniform (Figures 6.9–6.11; Tegmark et al. 2004; Abazajian et al. 2004) and it is certainly not static. Although the "cosmological principle" appears false,

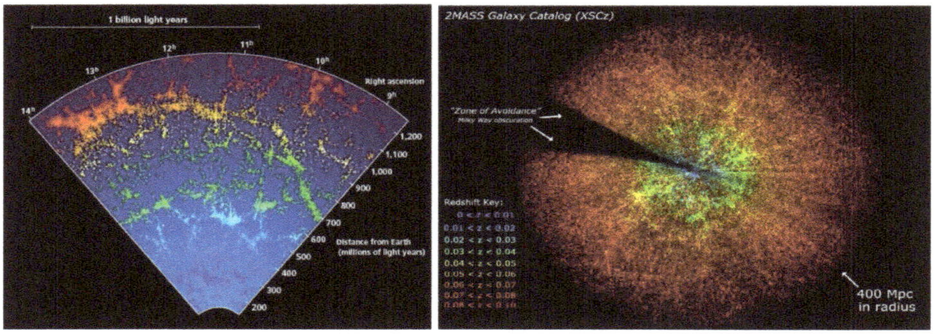

Figure 6.9. **Left (a):** On scales 10,000 times larger than galaxies we continue to see inhomogeneity. Each of the 9325 points in this image represents a galaxy for which the redshift (hence distance) has been measured, giving a 3D map. In addition to large chain-like structures that span the full region covered and are referred to as "walls" are large empty spaces or "voids." The full survey measured 205,443 galaxies providing the most accurate measurement of the density fluctuations in the universe. (Reproduced by courtesy of Sloan Digital Sky Survey (SDSS).) **Right (b):** 1.5 million galaxies from the 2Mass survey color-coded as a function of redshift. (Reproduced courtesy of SDSS.)

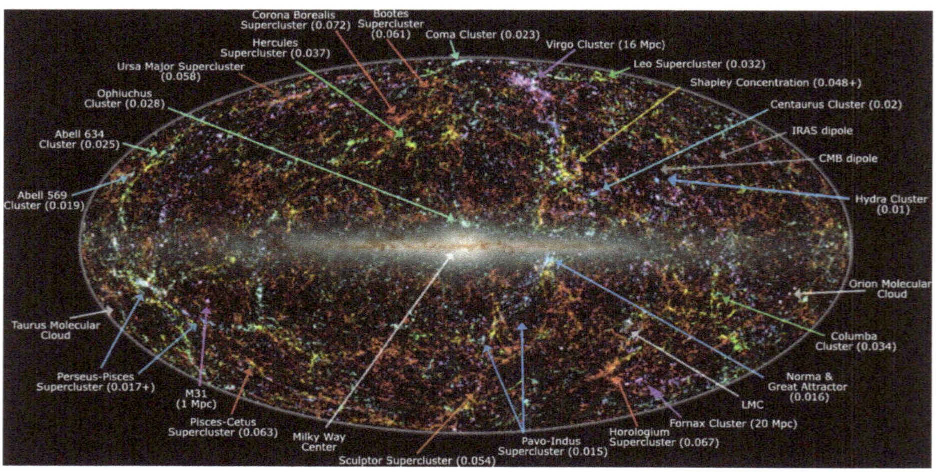

Figure 6.10. A view of the near-infrared sky showing the distribution of galaxies beyond the Milky Way. Data are from 2MASS Extended Source Catalog (XSC)—more than 1.5 million galaxies, and the Point Source Catalog (PSC)—nearly 0.5 billion Milky Way stars. The galaxies are color-coded by redshift (numbers in parentheses) obtained from the UGC, CfA, Tully NBGC, LCRS, 2dF, 6dFGS, and SDSS surveys (and from various observations compiled by the NASA Extragalactic Database), or photo-metrically deduced from the K band (2.2 μm). Blue/purple are the nearest sources ($z < 0.01$); green are at moderate distances ($0.01 < z < 0.04$), and red are the most distant sources that 2MASS resolves ($0.04 < z < 0.1$). The map is projected with an equal area Aitoff in the Galactic system (Milky Way at center). (Courtesy of NASA/Jarrett IPAC/Caltech).

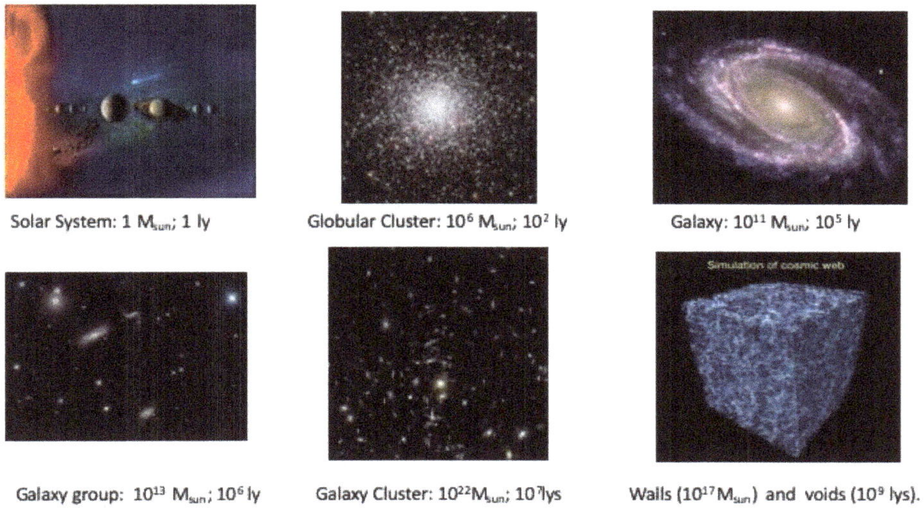

Figure 6.11. Cosmic Structure Hierarchy. Although we assume that the universe is homogeneous we see structure at the largest scales that we have observed. (Individual images reproduced courtesy of NASA.)

Einstein's theories explained many long-standing questions that Newton's theory of gravitation could not. Predictions made by Einstein's theories had been spectacularly confirmed with experiment (Eddington's observations of light deflection during a Solar eclipse in 1919 were considered to be conclusive proof of GR over Newtonian gravity; Eddington 1919). Systematic tests of GR starting in the 1960s with experiments within our solar system, and in the 1970s with observations of binary, pulsars found no deviations from the predictions of GR and hence our current picture of the universe is largely based on the gravitation theories developed by Einstein.

6.4 Expansion of the Universe

Einstein had found that solutions to his equations required a universe that was either expanding or slowing down; in order to "improve" his model to fit "observations" he added a cosmological constant, Λ, to his equations in order to keep the universe static. Not the first time that lack of adequate observations led a theoretician astray. However, analytic solutions to Einstein's equations formed by Alexander Friedmann were consistent with an expanding universe (Friedmann 1922, 1924).

6.4.1 Hubble's Law, Hubble Constant, Hubble Time

By the 1920s Hubble had found that all galaxies appeared to be moving away from us, and the farther away a galaxy was the faster it was moving away (Figure 6.12(a); Hubble 1929). The distance to the galaxies were measured spectroscopically: when

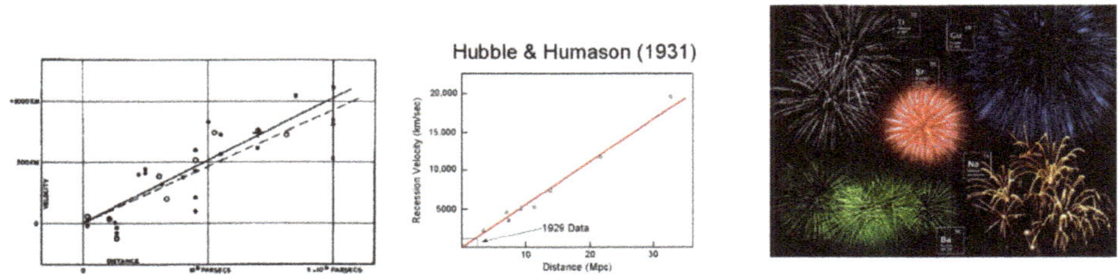

Figure 6.12. Left (a): Hubble's original result published in 1929: with no labels on the axes! Hubble's Law: the galaxies that are farther away from us are moving away faster than the nearby ones. (Reproduced courtesy of NASA.) **Center (b):** Hubble's result extended with observation by Humason (a more careful observer). (Reproduced courtesy of NASA.) **Right (c):** In fireworks each bloom of sparks has a single origin: the sparks that get farthest from the center are the ones that are moving fastest. Here as a bonus the elements that are used to make the different colors of the fireworks are noted. (Reproduced courtesy of NASA.)

an object is moving away from us, the light that it emits is "redshifted"[2] with the wavelengths getting longer. The amazingly linear relationship he observed was confirmed with further observations carried out by his assistant, Humason, and has since been named Hubble's Law (Figure 6.12(b); Hubble & Humason 1931). The relationship is somewhat like a fireworks display where the sparks that are farther from the center of the explosion are the ones that are moving faster (Figure 6.12(c)). In short, the universe was expanding. One obvious conclusion is that the galaxies were closer together at an earlier time, and at a time dependent on the rate of movement, they must have had a point of origin.

Hubble's law can be expressed very simply as v (velocity) = $H_o \times D$ (distance). H_o is called Hubble's constant. One of the key projects of the aptly named Hubble Space Telescope was to measure distant Cepheids to better determine the value of Hubble's constant (Freedman et al. 1994). H_o is expressed as the speed of a galaxy at a given distance from us and the currently favored value is 70 km/s/MPC (1 MPC, megaparsec, is 3×10^{19} km). The Hubble Time is the amount of time one predicts by projecting back assuming the universe has always been expanding at its current rate. This estimate of the age of the universe, 13.98 billion years, gotten by taking the inverse of the Hubble constant ($H_o = v/dD = 1/T$) is very close to the age of the oldest stars ($\approx$13.7 billion years), a fairly reasonable estimate. However most cosmology theories suggest that H_o is not constant. If the universe used to be expanding much faster than it does now, the Hubble time overestimates the age of the universe. If the universe is expanding faster today than it used to, the Hubble time underestimates the true age of the universe.

[2] This effect, called the Doppler effect, is an effect on light waves similar to what happens to sound waves emitted from a moving object. For example, if you are standing next to a railroad track and a train passes you while blowing its horn, you will hear the sound change from a higher (wave crests closer together) to a lower frequency (wave crests farther apart) as the train passes you by. As a result of this Doppler effect, at large redshifts, all of the ultraviolet and much of the visible light from distant sources is shifted into the infrared part of the spectrum by the time it reaches our telescopes.

6.5 The Need for Dark Matter

The motions of planets within our solar system are dictated by the force of gravity, which acts as the inverse square of the distance from the gravitational mass. The mass of the solar system is concentrated in the Sun and so the farther a planet is from the Sun the slower it moves (Figure 6.13(a)). In 1937, Fritz Zwicky, a Swiss astronomer, measured the motions of galaxies in the Coma cluster of galaxies (Zwicky 1937). He then determined the gravitational mass of the cluster based on the number of galaxies he counted in the cluster. Zwicky found that the velocities of the galaxies required a mass 400 times greater than that which he determined by counting. Although his measurements were off by an order of magnitude, his conclusion, that there must be matter he could not see causing the extra gravity, has held. Zwicky's contemporaries did not take his result seriously and it was more than 40 years before robust measurements by Vera Rubin verified the existence of "dark matter" (e.g., Rubin 1983). Rubin measured the velocities of stars and gas within edge-on spiral galaxies with a highly sensitive spectrograph. Her measurements showed that the velocities remained constant well beyond the regions containing the majority of the stars which represent the visible mass (Figure 6.13(b)). The galactic rotation curves show the velocities increasing as one accumulates more mass (represented by visible stars); but when one stops adding visible matter the velocities do not drop off as in our solar system. Rubin's work has been verified by many other astronomers and measurements of elliptical galaxies showed the same trend. Galaxy clusters are particularly useful as the need for dark matter can be proved in several independent ways: more accurate measurements of velocities in galaxy clusters; X-ray measurements of hot gas within the clusters (Figure 6.14); and from gravitational lensing effects on background objects (Figure 6.15).

The Wilkinson Microwave Anisotropy Probe (WMAP) determined that dark matter makes up 24% of the mass of the universe, and ordinary matter (atoms or baryons) make up less than 5%.

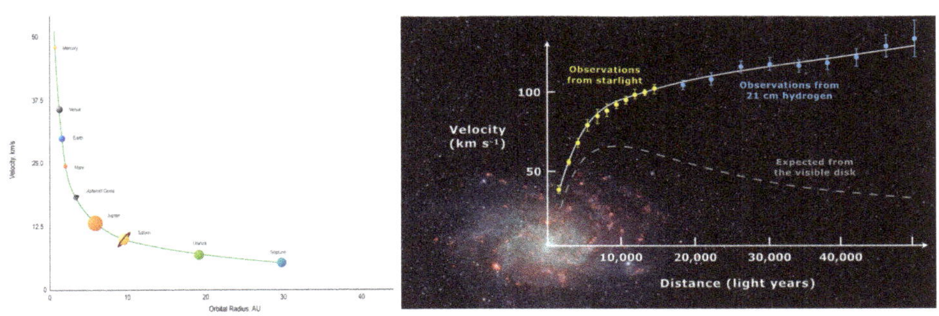

Figure 6.13. Left (a): The closer a planet is to the Sun the faster its velocity. (Reproduced courtesy of NASA/SSU/Aurore Simonnet.) **Right (b):** The velocity rotation curve of the galaxy M33 superposed on an optical image of the galaxy. From zero at the center of the galaxy the velocity increases as more and more mass is encompassed and levels off at a high rate (yellow and blue points). The dashed white line shows the expected behavior for the matter (stars) that is observed. (Reproduced courtesy of Mario De Leo.)

Figure 6.14. The "Bullet Cluster," 1E 0657-56 is a result of the collision of two separate clusters. The image is a composite of optical data from HST and the Magellan telescopes in Chile; these show the galaxies in white and orange. The red clumps are the hot intercluster gas seen in X-rays with the Chandra X-ray Telescope. Superposed in blue is the dark matter inferred from the gravitational lensing seen on the galaxies. It is clear that the colliding gas is slowed down by the collision. The galaxies and the dark matter associated with them pass through relatively unchanged. (Reproduced courtesy of X-ray: NASA/CXC/CfA/M. Markevitch et al.; Optical: NASA/STScI; Magellan/U. Arizona/D. Clowe et al.; Lensing Map: NASA/STScI; ESO WFI; Magellan/U. Arizona/D. Clowe et al.)

Figure 6.15. Left (a): An excellent example of gravitational lensing. Galaxy cluster Abell 2218 has enough mass that multiple distant objects appear as arcs in the image. The visible mass alone cannot provide the required mass to create such distortions in spacetime. The only solution is the introduction of dark matter. (Reproduced courtesy of NASA, ESA, and Johan Richard (Caltech, USA). Acknowledgement: Davide de Martin & James Long (ESA/Hubble). **Right (b):** Superposed on this Hubble Advanced Camera for Surveys image of the galaxy cluster CL 0024+17 is a map (in blue) that shows the clusters dark matter distribution as derived from how the light of more distant galaxies is distorted by gravitational lensing. (Reproduced courtesy of NASA, ESA, M.J. Jee and H. Ford (Johns Hopkins University).)

6.6 The Need for Dark Energy: Acceleration of the Universe

In 1998, two separate projects studying supernova Type 1a—used as a standard candle because they have similar intrinsic brightness—found that some of these objects appeared fainter than they should be considering the distance to their host galaxies (Reiss et al. 1998; Perlmutter et al. 1998). The distances to the supernovae are measured by two independent means: their apparent luminosity using photometric observations, and their redshifts using spectroscopic observations. The redshift is a measure of the extent to which the universe has expanded relative to the time the supernova occurred. The redshift distances measured by both groups gave distances that were too near for the faintness of the detected supernovae. To explain this observation, they suggested that the expansion of the universe is actually accelerating, or speeding up. This was perplexing, because the gravity inherent in dark matter should have been strong enough to prevent such an expansion. This suggested the existence of some other material, something with an antigravity effect, causing the accelerated expansion of the universe. That extra material is now called dark energy.

In Figure 6.16, Ω_Λ represents dark energy and Ω_M matter density. The supernova data have a large range of uncertainty, but studies of the CMB also showed the need for dark energy with lesser uncertainty, and studies of clusters of galaxies yielded a good measurement of the matter content but were ambiguous on dark energy content of the universe (Suzuki et al. 2012). However the different measurements all

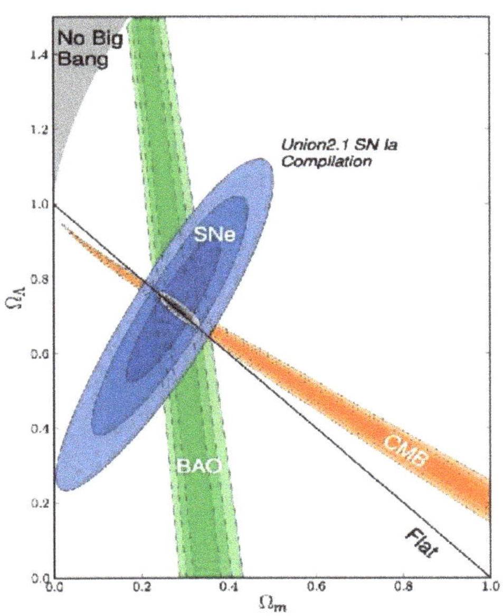

Figure 6.16. A schematic showing the relative contributions of matter and dark energy measured with three techniques. The blue ellipses show values obtained from studies of SN Ia; with the outermost ellipse showing the greatest certainty; the green shows the results from a specific galaxy cluster survey (BAO), and the orange results from the CMB results of WMAP and Planck. (Suzuki et al. 2012). (Reproduced courtesy of Saul Perlmutter/David Rubin.)

converged on a fairly narrow determination: 30% matter and 70% dark energy. Note that dark energy is identified with the symbol, Λ. Λ was Einstein's fudge factor but now, instead of being used to make the universe static, it is used to accelerate the expansion of the universe!

6.7 Standard Model

The "standard" model of the universe[3] takes general relativity as the correct theory of gravity, accepts the presence of dark matter and dark energy, and assumes that the "Big Bang" that started the universe was caused by pure energy. Although not perfect, it is the simplest model that explains currently observed properties of the universe. Called variously the Standard Model of Cosmology (SMC), the Lambda Cold Dark Matter (ΛCDM, or Lambda-CDM) model, where Λ, the cosmological constant, represents dark energy at about 71% of the total mass-energy of the Universe, the dark matter (about 24% of the total mass-energy); it is assumed to be "cold" because it has to be slow compared to the speed of light in order to stay within the potential wells of galaxies (Blumenthal et al. 1984; Peebles 1982); ordinary matter is less than 5% of the total mass-energy. ΛCDM is able to explain the large-scale distribution of galaxies (Figures 6.9, 6.10, and 6.11 the CMB and its detailed substructure (Figures 6.18, 6.19, and 6.20), as well as the observed abundances of hydrogen, helium, and lithium (Figure 6.21), and the accelerating expansion of the universe.

6.8 The Beginning of the Universe

Einstein knew that the weak, strong, and electromagnetic forces worked on relatively short distances so he assumed only gravity acts over long distances. Einstein's General Relativity theory required space and time to be defined by the matter and energy content of the universe and implied that there was a time when all matter and energy were combined in one singular point. Our current working model of the universe suggests that this point where everything began existed roughly 13.7 billion years ago (Figure 6.17). Using known physical laws to extrapolate back from now to the beginning of our universe, shows the start as a state of immense density and temperature (Weinberg 1977). In fact, temperatures were so high (greater than 2967° Kelvin (4880°F)) the very atoms that generate the electromagnetic radiation we perceive could not form. Neutrons and protons existed in roughly equal amounts since the energy was high enough to exchange them freely back and forth. The universe was essentially a hot, dense, plasma. By about 380,000 years after the Big Bang, the universe had cooled down to about 3000K, a temperature at which electrons could combine with protons and form atoms. This time is called recombination, although it is actually the first time that electrons and

[3] Not to be confused with the Standard Model of particle physics (SMPP) which combines the electromagnetic, weak, and strong forces to describe all currently known elementary particles. SMPP does not use Einstein's theory of gravity and cannot explain the accelerating expansion of the universe. While the SMC incorporates the SMPP, neither model has a viable candidate for dark matter.

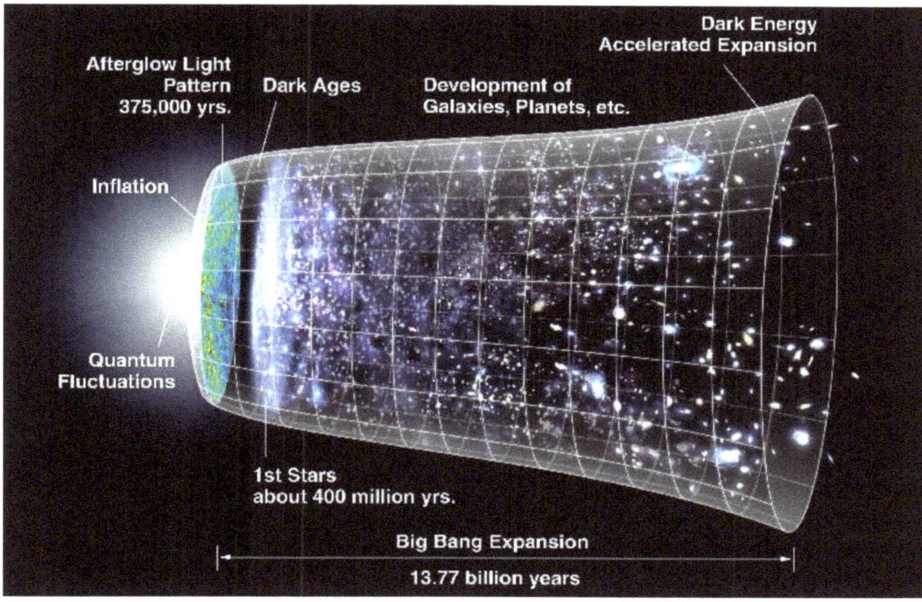

Figure 6.17. A schematic depicting the history of the universe from the Big Bang. Specific events are shown such as the photons we now call the Cosmic Microwave Background at 375,000 years post Big Bang. After this the constant interaction between matter (electrons) and light (photons) made the universe opaque (Dark Ages). At about 400 million years after the Big Bang the universe had cooled enough that protons could capture electrons and the formation of neutral hydrogen made the first stars possible. (Reproduced courtesy of NASA.)

protons combine. Only the lightest atoms formed at this time: mostly hydrogen (about 75% by mass), helium (about 25% by mass), and trace amounts of lithium. Helium atoms have a larger binding energy than the next heaviest elements and so prevented their formation at this time. Once atoms were formed the universe became transparent because the atoms could emit photons that moved freely through space. This is referred to as the point of last scatter.

6.8.1 The Cosmic Microwave Background

Amazingly, we have detected the radiation from the point of last scatter. The first hint came in the 1960s when Arno Penzias and Robert Wilson of Bell Labs (while checking for possible interference for communication channels at certain wavelengths) found an unexplained signal that persisted regardless of direction in the sky or time of day or year (Penzias & Wilson 1965). The signal they observed at a temperature of about 3K was consistent with that predicted by Robert Dicke and collaborators (Dicke et al. 1965) as the expected remains of the initial "Big Bang" that started our universe; earlier predictions by Alpher and Herman in 1948–1950 had miscalculated the expected temperature (Alpher & Herman 1948). Dicke had just begun to design a radiometer to search for the effect when he heard about Penzias and Wilson's discovery and realized they had already detected the signal. Confirmation of the detection came with instruments flown on NASA's Cosmic

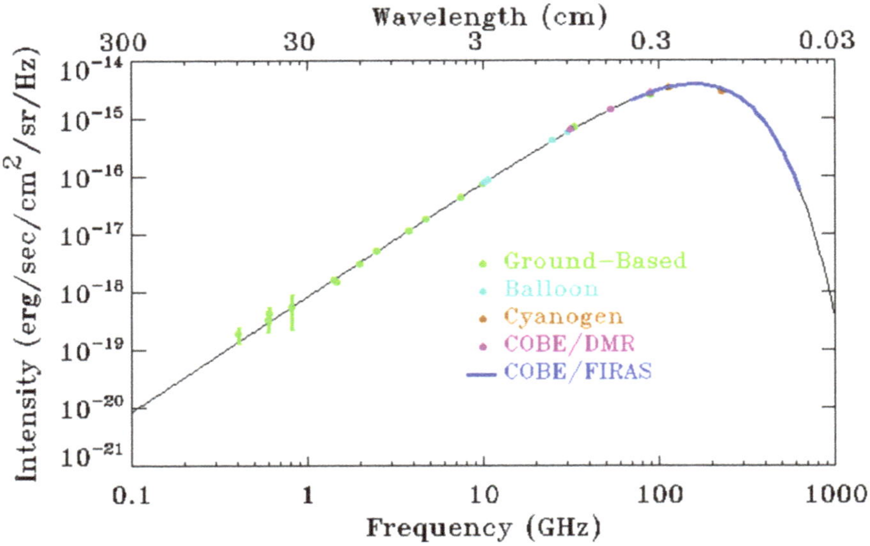

Figure 6.18. Measurements of the intensity of the cosmic microwave background taken with instruments from the ground, on balloons, and from space provides information as a function of observing frequency (or wavelength). The original measurements by Penzias & Wilson were at a frequency of 4 GHz. The CMB follows the expected curve of a 3 degree blackbody over more than 5 orders of magnitude in intensity. (Reproduced courtesy of NASA/GSFC.)

Background Explorer (COBE) satellite launched in 1989. The Far-InfraRed Absolute Spectrophotometer (FIRAS) on COBE led by Mather et al. (1990) showed that the spectrum of the microwave background was a perfect blackbody at a temperature of 2.728K (Figure 6.18). In January 1990, John Mather received a prolonged standing ovation from an audience of over 2000 Astronomers when he presented this result at the 175th American Astronomical Society meeting.[4] COBE also carried the Differential Microwave Radiometer (DMR; Smoot et al. 1991) that measured the minute variations of the temperature over the sky (Figure 6.19(b)). The DMR was limited to measure scales down to about 7 degrees. Succeeding satellites, NASA's Wilkinson Microwave Anisotropy Probe (WMAP) and ESA's Planck mission, refined these to scales down to 1 degree (Figure 6.19(c); Bennett et al. 2003) and a tenth of a degree (Figure 6.19(d); Aghanim et al. 2020).

If one plots the temperature fluctuations as a function of angular scale on the sky a very interesting oscillation pattern emerges as one is able to go to finer and finer angular scales (Figure 6.20; Hinshaw et al. 2003a, 2003b). These are referred to as acoustic oscillations: gravity's attempt to compress the initial plasma fluid is resisted by the radiation pressure of the photons, resulting in oscillations. Compressing a gas heats it, expanding a gas cools it. The "acoustic" peaks are higher temperatures which imply higher densities of matter, the troughs are regions of lower density.

[4] The author was in the audience and found it a profoundly uplifting moment.

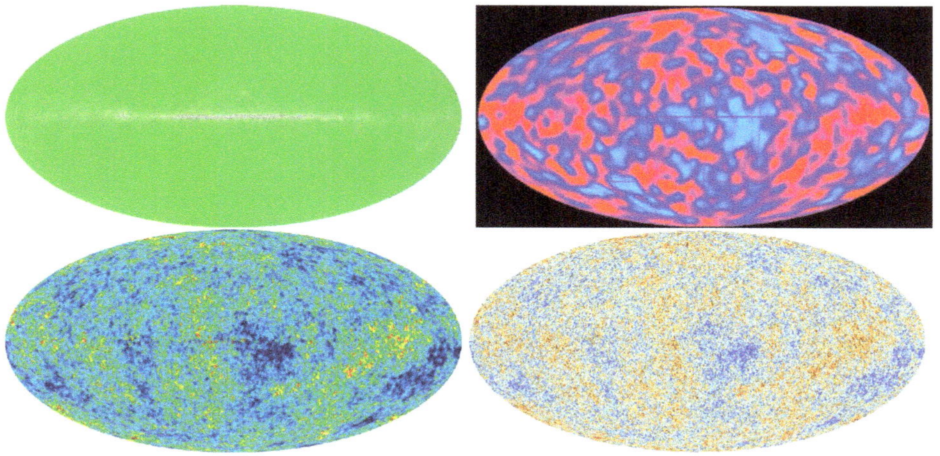

Figure 6.19. Upper left (a): The original detection of the Cosmic Microwave Background was made by Arno Penzias and Robert Wilson in 1965 using Bell Lab's Holmdel Horn Antenna radio telescope. They found emission at a temperature of 3K coming from every direction in which they looked (green in image). This is a simulation of their findings. The emission in white is contamination from the Milky Way Galaxy. (Reproduced courtesy of NASA/WMAP.) **Upper right (b):** Twenty five years later, the Differential Microwave Radiometer on NASA's COBE satellite was able to show temperature fluctuations of parts in a hundred thousand at the limiting scale of 7 spatial degrees on the sky (Mather et al. 1990). (Courtesy of NASA.) **Lower left (c):** In 2003 with one year of observations with the Wilkinson Microwave Anisotropy Probe (WMAP) showed fluctuations of order parts per million down to scales of 1 spatial degree. Red depicts higher temperatures and blue cooler temperatures (the full range is only ±200 microKelvin; Bennet et al. 2003). (Courtesy of NASA/WMAP Team.) **Lower right (d):** Europe's Planck mission launched in 2009 measured temperature fluctuations down to a tenth of a spatial degree at multiple frequencies showing the very fine differences in the temperature distribution that likely led to the large-scale structure we see today (Aghanim et al. 2020). (Reproduced courtesy of NASA/Planck.)

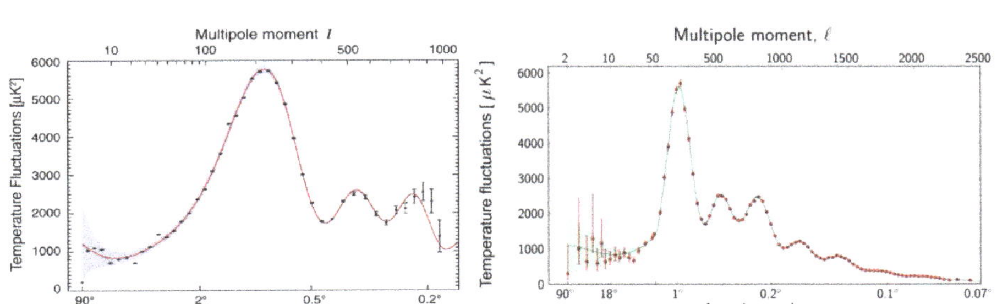

Figure 6.20. Left (a): A 5 year power spectrum of the CMB as measured by WMAP together shows a dominant peak at 1° scales. The red line is a theoretical calculation. (Reproduced courtesy of NASA/WMAP Science Team.) **Right (b):** The spectrum as measured by Planck showed 7 peaks each relatable to a physical measurement. The green line is a theoretical calculation. (Reproduced courtesy of ESA and the Planck Collaboration.)

At the point where the temperature is cool enough for electrons and protons to combine (recombination) the oscillations are frozen. The temperature fluctuations in the figure are measured in millionths of a degree Kelvin. As the universe expands the wavelengths of CMB photons stretch and the temperature of the CMB drops. The increase in wavelength means a decrease in energy. This effect is referred to as the cosmological redshift. So a doubling in the size of the universe means a halving of the temperature of the CMB.

COBE established that there are fluctuations at the largest angular scales (Figure 6.20). At these scales we see the initial conditions of the universe. WMAP distinguished three peaks. The first peak at slightly less than 1 degree is a measure of the curvature of the universe and is consistent with a flat universe, the second peak is a measure of the amount of baryons. Consider the analogy of a spring: if you add a load to the spring it will oscillate up and down; the height of the oscillation is related to the mass of the baryons. The third peak is a measure of the dark matter.

The fluctuations of the CMB are determined by the composition, density, and expansion rate of the universe. Hence the composition, density, and expansion rate of the universe can be measured by studying the fluctuations of the CMB. This is exactly what WMAP and Planck did. WMAP measured temperature fluctuations to one part in 100,000 (due to gravitational redshifting by pockets of higher density), estimated the age of the universe to be 13.772 ± 0.043 billion years. The Planck Collaboration estimated the age of the universe to be 13.813 ± 0.038 billion years. The two are in agreement within measurement errors.

6.8.2 The First Stars and Galaxies

The oldest (i.e., the earliest formed) galaxies and quasars we have observed, date from about a billion years after the Big Bang. After the burst of light that generated the CMB when atoms first formed there was a long period of darkness (Figure 6.17). Cosmological models predict that this is the amount of time it took for those tiny fluctuations in density detected in the CMB to grow into galaxies (Rees & Sciama 1969). Clearly stars must have formed before the galaxies. According to cosmological models, the first stars should have appeared somewhere between 200 million and 400 million years after the Big Bang and were composed of the hydrogen and helium atoms created in the Big Bang. Simulations of star formation using only hydrogen and helium particles show that the first stars were of greater mass and rotated much faster than the stars we see today (Larson & Bromm 2001). It is not until the first stars died releasing heavier elements, that stars like our Sun could be born (Figure 6.21).

6.9 The Physical Limits and Contents of the Universe

Figure 6.22 locates many kinds of objects observable in the known universe all on one plot. It is a logarithmic plot because it covers many orders of magnitude. On the y-axis we have mass in kilograms ranging from 10^{-70} to 10^{56} kg and on the x-axis we have size in meters ranging from 10^{-35} to 10^{30} m. The physical limits on the observable universe are set by the purple, black, and red lines. The red line denotes a distance of 10^{26} m, beyond this distance light has not yet had time to reach us. The

Figure 6.21. Left (a): The first generation of stars, also known as Population III stars, were made almost entirely of hydrogen and helium. Later generations of stars, including the Sun, contain heavier elements like oxygen, carbon, and iron, which formed in the cores of earlier stars. (Reproduced courtesy of NASA/WEBB.) **Right (b):** An image of the galaxy GN-z11 observed with HST in the Great Observatories Origins Deep Survey (GOODS). The fields shows tens of thousands of galaxies. GN-z11 is one of the most distant galaxies seen at 13.4 billion years in the past. The optical light appears red because of the expansion of the universe; the light is actually from young, bright blue stars. (Reproduced courtesy of NASA, ESA-Hubble, P. Oesch (Yale University), G. Brammer (STScI), P. van Dokkum (Yale University), and G. Illingworth (University of California, Santa Cruz.)

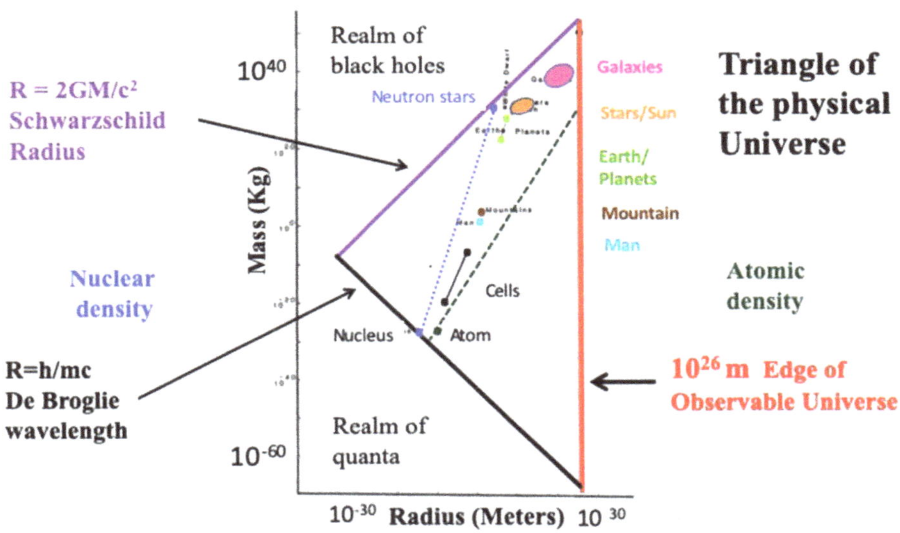

Figure 6.22. If we plot mass versus radius of all known objects in the universe we define its physical limits. Above the purple line is matter whose density is so high that light cannot escape it (and hence we cannot get information from it): the realm of black holes. Below the black line we cannot determine both energy and speed: the realm of quanta. The red line designates the edge of the known universe. (Adapted from a schematic sketched by Paul Doherty, 1947–2017).

purple limit is defined by the radius of the event horizon of a black hole. For a given radius if mass is above this line it defines a black hole and we can have no communication with the inside of a black hole so it cannot be known. Finally the black line represents the de Broglie wavelength for matter going at the speed of light. If the velocity is at the speed of light than anything below here is in the realm of quantum physics and cannot be determined. Everything that we could have actually seen or measured is contained within these limits. But in fact, everything we have seen or measured falls within a much smaller area. The blue dotted line shows matter at nuclear density, the end points show a hydrogen nucleus at a radius of 10^{-15} m with a mass of 10^{-27} kg and a neutron star with a radius of 10km and a mass of 5×10^{30} kg. The green dashed line shows atomic density; everything we know physically lies between these two lines: cells occupy a large range in this space from viruses to amoeba. But the mass/radius range of all humans lies within one dot. Relatively speaking the allowed ranges of stars and galaxies, as given by the circles here are quite small. And it is clear that if something the size of a nucleus has a mass of 10^{20} kg it is a black hole: density greater than nuclear density.

6.10 The End of the Universe

The total amounts and densities of matter, dark matter and dark energy in the universe play crucial roles in determining the geometry of space. They also determine the final fate of the universe. WMAP showed that the curvature of space was flat to within 4% (Figure 6.23). Effectively by measuring large enough triangles. Then Ω_{Total} = density/critical density = 1. This is consistent with the Inflationary Theory, an extension of the Big Bang theory which predicts that density is very close to the critical density, producing a flat universe.

After the WMAP results and the SN1a studies requiring acceleration it became clear that the content of the universe included both matter and dark energy (Figure 6.16(a)). The matter content, Ω_M = regular matter + dark matter, is less than 30% of the total. The dark energy content, Ω_Λ is greater than 70% of the total. Then $\Omega_{\text{Total}} = \Omega_M + \Omega_\Lambda$.

Before the need for dark energy, the shape of the universe was determined entirely by its matter density. The total matter density Ω_M (visible matter + dark matter) was the total density. If the density is low the universe expands forever; if the density is high the universe will eventually collapse under its own gravity (Figure 6.24).

6.10.1 Things fall apart…

Will the universe expand forever? In a matter dominated universe where we are now, atoms, people, and galaxies stay together because of gravity. In a dark-energy dominated universe, even the atoms in our bodies will start to expand and be ripped apart (things fall apart…). Since the action of gravity works against expansion, if the density is large enough, the expansion would eventually stop and the universe would collapse: this is called a closed universe. Closed means gravity wins and things will eventually fall back. However, it is possible that the fall back will cause such a crush that we will have another Big Bang and everything will start all over. If the density is small

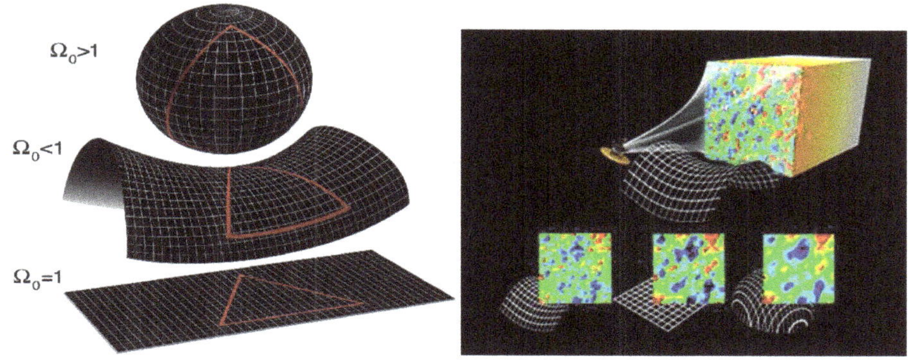

Figure 6.23. **Left (a):** The local geometry of the universe is determined by whether the density parameter, Ω, is greater than, less than, or equal to 1. If $\Omega > 1$, the "shape" is spherical and the sum of the angles of a triangle drawn on its surface will add up to greater than 180 degrees (Riemannian universe); this results in a "closed" universe. If you travel long enough in one direction you will come back to the starting point. If $\Omega < 1$ the shape is like a saddle and the sum of the angles of a triangle drawn on its surface will add up to greater than 180 degrees (Lobachevskian universe); This results in an "open" universe which is infinite. If $\Omega = 1$ we have a flat universe and the sum of the angles of a triangle drawn on its surface will add up to exactly 180 degrees (Euclidean universe); a flat universe is also infinite. (Reproduced courtesy of NASA/WEBB.) **Right (b):** This animation shows that WMAP results are consistent with a flat universe. (Reproduced courtesy of NASA/WMAP.) Animation available at http://doi.org/10.1088/2514-3433/ac81b3.

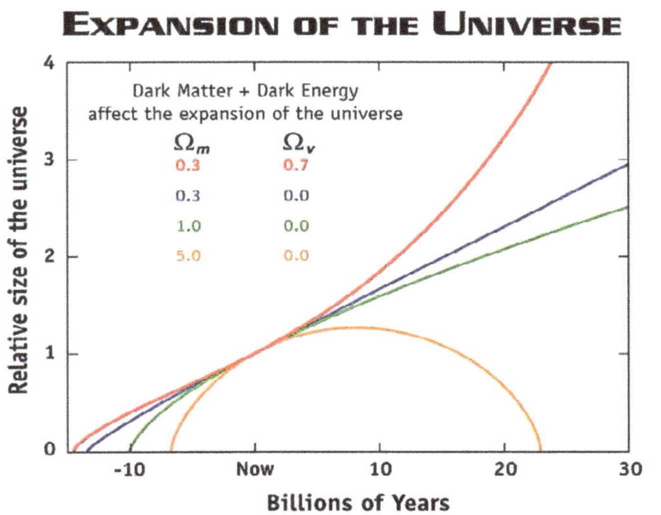

Figure 6.24. A schematic showing possible pasts and futures of the universe. At the present moment we appear to be at a unique location consistent with all possible models. The ΛCDM (red) shows an accelerating universe. (Reproduced courtesy of NASA/WEBB.)

enough, the expansion would continue forever (an open universe). At a certain precise critical density (9×10^{-27} kg m^{-3}), roughly five hydrogen atoms per cubic meter, the universe would asymptotically approach zero expansion rate, but never collapse.

6.10.2 The Center Cannot Hold…

Although the fireworks analogy works well to explain the observed relationship of galaxy distances to their velocities it is completely inadequate to explain the point of origin. Consider the light that reaches us from the Sun: we know that light has a finite speed and we know the distance to our Sun and we can calculate that it takes light a bit over 8 minutes to get to us from the Sun. So when we are looking at the Sun we are not only seeing an object that is 14.9e7 km (93 million miles) away we are also looking 8 minutes in its past. The star Alpha Centauri is about 3.8e13 km (4 light-years) from us, the light we see from Alpha Centauri is 24 billion miles away and at any instant we are seeing it as it was 4 years ago. The farther away an object is the greater the distance it travels to us and the longer into the past that we are looking. So what happens if are able to you look at an object that is so far away that it took light from it a time equivalent to the age of the universe to reach us? If we looked for such objects in every direction, we are looking as far away as we can, but we are also looking at the point from which everything is expanding, that is when everything was all together. But that all together is the size of the universe: there is no center; the "beginning" occurred everywhere. A common analogy is the surface of a balloon: think of the surface of the balloon as defining the universe, paste some dots on it to represent galaxies. If you then blow up the balloon, every galaxy is moving away from every other galaxy, each galaxy stays the same size, and there is no center. Of course, this analogy uses a two-dimensional universe. Note that the center of the balloon is actually in a 3rd dimension. As we learned in Chapter 1, in Einstein's GR it is space itself that is expanding. However, it can be said that everyone is at the center of their own universe (Figures 6.25 and 6.26); somewhat like Navajo cosmology as best as I understand it:

Figure 6.25. This animation shows simulations of galaxies, gas, and dark matter including the effects of gravity and hydrodynamics. Beginning 20 million years after the Big Bang (13.7 billion years ago) to present day. (Reproduced courtesy of NASA/Cen & Ostriker, AVL (http://avl.ncsa.illinois.edu/) at NCSA (http://ncsa.illinois.edu/), University of Illinois (http://www.illinois.edu).) Animation available at http://doi.org/10.1088/2514-3433/ac81b3.

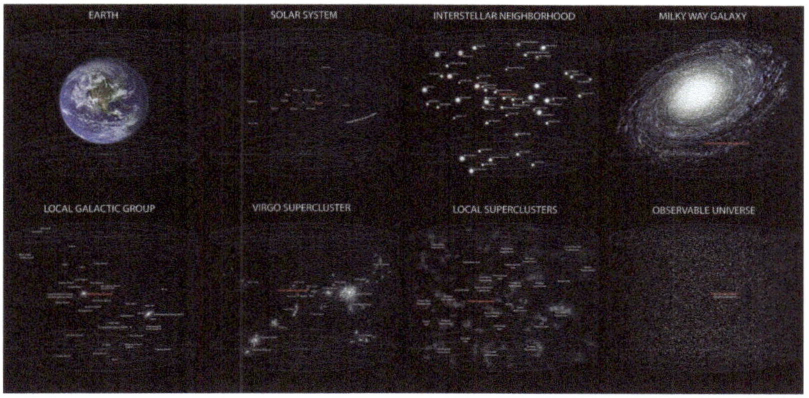

Figure 6.26. The universe, showing the solar system, the nearest stars, the Milky Way, the Local Group (nearby galaxies), the Lanikia Supercluster (which includes the Virgo Supercluster), the nearest superclusters, and the observable Universe for which there is no specific center. (Reproduced courtesy of Andrew Z. Colvin.)

every hogan is the center of its own universe. It is a pleasing concept but also makes clear that no two people see the same universe.

References

Abazajian, K., Adelman-McCarthy, J. K., Agüeros, M. A., et al. 2004, AJ, 128, 502

Alpher, R. A., & Herman, R. 1948, Natur, 162, 774

Bennett, C. L., Halpern, M., Hinshaw, G., et al. 2003, ApJS, 148, 1

Blumenthal, G. R., Faber, S. M., Primack, J. R., & Rees, J. 1984, Natur, 311, 517

Dicke, R. H., Peebles, P. J. E., Roll, P. G., & Wilkinson, D. T. 1965, ApJ, 142, 414

Eddington, A. S. 1919, Obs, 42, 119

Friedmann, A. 1922, ZPhy, 10, 377

Friedmann, A. 1924, ZPhy, 21, 326

Freedman, W. 1994, ApJ, 427, 628

Hinshaw, G., Barnes, C., Bennett, C. L., et al. 2003a, ApJS, 148, 63

Hinshaw, G., Spergel, D. N., Verde, L., et al. 2003b, ApJS, 148, 135

Hubble, E. 1925, PA, 33, 252

Hubble, E. 1929, PNAS, 15, 168

Hubble, E., & Humason, M. L. 1931, ApJ, 74, 43

Larson, R. B., & Bromm, V. 2001, SciAm, 285, 64

Leavitt, H. S. 1908, AnHar, 60, 87

Leavitt, H. S., & Pickering, E. C. 1912, HarCi, 173, 1

Mather, J., Cheng, E. S., Eplee Jr, R. E., et al. 1990, ApJ, 354, 37

Peebles, P. J. E. 1982, ApJ, 263, L1

Penzias, A. A., & Wilson, R. W. 1965, ApJ, 142, L419

Perlmutter, S., Aldering, G., Della Valle, M., et al. 1998, Natur, 391, 51

Planck Collaboration, Aghanim, N, Akrami, Y., et al. 2020, A&A, 641, A1

Rees, M. J., & Sciama, D. W. 1969, CoASP, 1, 153

Riess, A., Filippenko, A. V., Challis, P., et al. 1998, AJ, 116, 1009

Rubin, V. 1983, Sci, 220, 1339
Shapley, H. 1917, ApJ, 46, 64
Shapley, H. 1918, ApJ, 48, 154
Smoot, G. F., Bennett, C. L., Kogut, A., et al. 1991, AdSpR, 11, 192
Suzuki, N., Rubin, D., Lidman, C., et al. 2012, ApJ, 746, 85
Tegmark, M., Blanton, M. R., Strauss, M. A., et al. 2004, ApJ, 606, 702
Weinberg, S. 1977, The First Three Minutes (2nd ed., 1993, New York: Basic Books)
Zwicky, F. 1937, ApJ, 86, 217

Chapter 7

The Ever Expanding Horizon

Colorful universe. Reproduced courtesy of NASA.

7.1 Introduction

Space and ground-based astronomical instruments have covered essentially the entire electromagnetic spectrum with ever increasing sensitivity and resolution. In addition new channels for gathering information (cosmic rays, neutrinos, gravitational waves) have been brought into play. Every addition of wavelength range and channel, any significant progress in sensitivity and resolution (angular, spectral, or timing) in a given area has led to tremendous increases in our understanding. They have also led to new questions and often to discoveries of entirely new classes of objects. Because they

doi:10.1088/2514-3433/ac81b3ch7

are not stopped by intervening matter and magnetic fields gravitational waves and neutrinos allow us to look far farther than possible with electromagnetic waves or cosmic rays. All three non-electromagnetic channels appear to be associated with violent events involving high masses and very high energies that have implications for fundamental theories of physics (nuclear and particle), and gravity.

The advances in computational power over the past fifty years have also had a revolutionary impact on astronomy. High-performance and high-throughput computing are essential for detailed simulations of actual physical processes and for analysis of increasingly large data sets. Computer intensive processes include interferometry which can be used to achieve significant increases in angular resolution, and adaptive optics which can be used by large ground-based optical telescopes to alleviate the effects of atmospheric turbulence. Super computers are needed to deal with the petabytes of data that can now be gathered.

In this chapter I will highlight a few of the many current state-of-the-art instruments, some recently completed instruments, and several instruments and techniques that are currently under development. In operation over the next 10–50 years these will answer many of the questions from the past 10–50 years and, without doubt, raise new questions and add new complexities to our understanding of the universe.

7.2 Electromagnetic Waves

7.2.1 Radio

Radio telescopes can penetrate the dust and gas of the interstellar medium much better than optical telescopes because they operate at much longer wavelengths. Radio astronomers were forced to use interferometry early on. Since angular resolution increases as the square of the aperture and decreases with wavelength, long radio wavelengths mean that radio telescopes must be much larger than optical telescopes to achieve the same angular resolution. Interferometry, the technique of combining light from multiple telescopes to increase angular resolution and light gathering capacity, requires significant computing power in order to add the signals from multiple telescopes coherently, i.e., in a way that preserves the phase information in the incoming radiation. The effective angular resolution depends on how far apart the individual elements are spaced and the amount of light gathered increases directly as the number and size of the elements used.

7.2.1.1 The Square Kilometer Array (SKA)

Because the efficiency and effectiveness of radio telescopes can be increased by increasing the number of telescope elements, a large international collaboration (14 member countries and participation from over 100 organizations in 20 countries) was organized to develop a telescope that is more than an order of magnitude greater in sensitivity than existing facilities at meter to centimeter wavelengths.

The SKA started construction in 2013 and expects to be completed in the late 2020s. At a frequency range of 110 mHz to 12.5 GHz with a telescope covering a million square meters of collecting area, SKA will provide 50 times the sensitivity of the Very Large Array (VLA). If placed at a distance of 500 light-years, our own solar system would be about 1 arcsec across, so observations at milliarcsec angular resolution are required to zoom in on planets. The SKA (Figures 7.1 and 7.2) will have this resolution. When operating at its highest frequencies, the SKA will be able to image the thermal emission from dust in the "habitable zone" around other stars in unprecedented detail. This will allow identification of features in disks related to planet formation, e.g., nearly empty gaps and spiral waves, track the evolution of these features, and identify rapid gravitational instabilities. Moreover, the telescope will be so sensitive that it will be able to detect signals comparable in strength to television transmitters operating on planets around the closest stars to the Sun. The SKA will accordingly be able to search for signals from other civilizations in a much larger volume than has been possible so far.

Figure 7.1. Left (a): The Square Kilometer Array (SKA) will have hundreds and eventually thousands of mid to high frequency 15 m dishes located in South Africa and Africa. (Reproduced courtesy of the SKA Project Development Office and Swinburne Astronomy Productions.) **Right (b):** Hundreds of thousands and eventually up to a million low-frequency antennas will be located in Western Australia. (Reproduced courtesy of the SKA Project Development Office and Swinburne Astronomy Productions.)

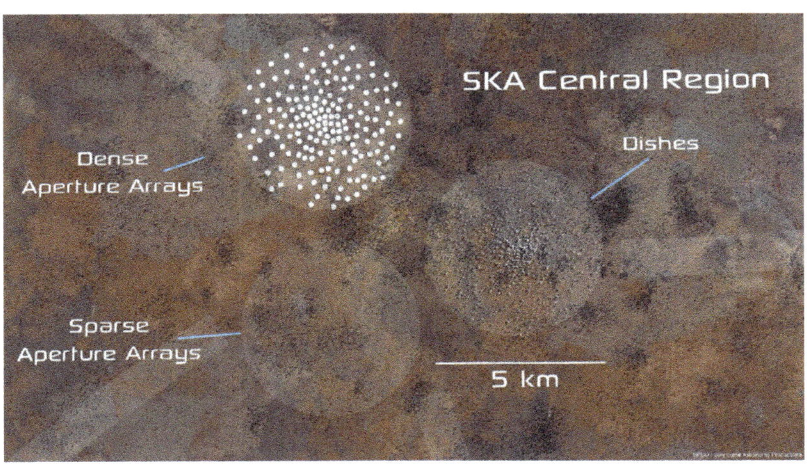

Figure 7.2. An aerial overview of the central region of the SKA with the three 5 km diameter cores of Antennas, Dense Aperture Arrays, and Sparse Aperture Arrays. (Reproduced courtesy of the SKA Project Development Office and Swinburne Astronomy Productions.)

7.2.1.2 Next Generation Very Large Array (ngVLA)

The capabilities being developed for ngVLA will result in an observatory that will revolutionize many areas of astrophysics. Covering the frequency range from 1.2 to 116 GHz (0.26–25 cm) the ngVLA will fill in the wavelength range between ALMA (84–720 GHz; sub-mm wavelengths; see Chapter 1) and SKA (110 mHz–12.5 GHz; decimetric wavelengths) at the same level of sensitivity and resolution (an order of magnitude above that of the VLA). ngVLA will comprise two hundred and forty-four 18 m diameter and nineteen 6 m diameter dishes. Designs for the 18 m telescopes are currently underway and are expected to be operational by 2024. The 6 m dishes will comprise a Short Baseline Array (SBA; shown in green in Figure 7.3). Of the 18 m dishes, thirty (the Long Baseline Array, LBA) will encompass baselines up to 9000 km (Figure 7.3).

ngVLA's frequencies of operation will allow it to study the dust emission of protoplanetary disks that is not feasible with any other facility (at the frequencies ALMA is sensitive to, the dust continuum is optically thick). This is the region where Earth like planets form. One aim is to measure planets orbiting their suns as close as 0.5 AU (1.5×10^7 km) up to a distance of 140pc (4.3×10^{15} km) from us. The resolution of ngVLA will enable it to map circumplanetary disks and mm-size dust particles in order to understand the formation of planetary systems like our own (Figure 7.4(a)). ngVLA will also be able to detect complex molecules that are critical for star formation and possibly important as building blocks for life, leading to a better understanding of the habitability of exoplanets (Figure 7.4(b)). Another goal is to determine the amount of "fuel" in the form of cold gas that is available for forming stars in early galaxies and forming galaxies through the age of the universe. ngVLA will also detect, at fine resolution, pulsars in the Galactic center, enabling new tests of gravity theories. It will be able to separate low-luminosity nearby black holes from background sources and to identify the radio counterparts of transient sources discovered at other wavelengths.

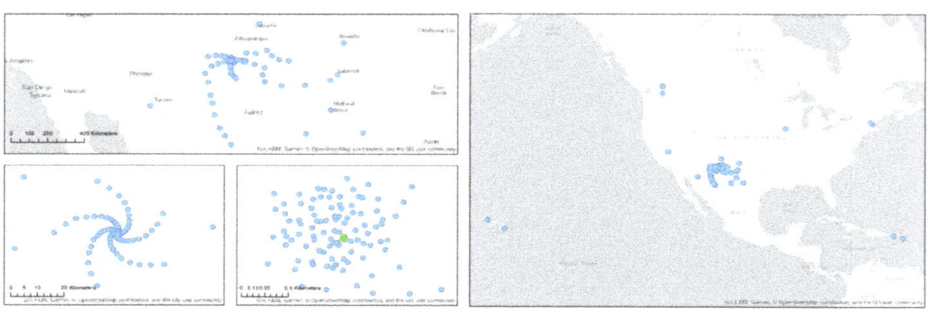

Figure 7.3. The Main Array Configuration (MAC) of 214 eighteen meter dishes and the Short Baseline Array (SBA in green; 19 six meter dishes) of the ngVLA. The MAC achieves high surface brightness sensitivity and high fidelity imaging on angular scales from 1000 to 1 mas by having a large collecting area fraction randomly distributed in the compact array core, spiral arms extending out to 36 km for snapshot imaging, and Long Baseline Array (LBA) extending asymmetrically to 1000 km baselines to fill the u,v plane with Earth rotation and frequency synthesis. A LBA provide continental-scale extended baselines (8860 km), while the SBA recovers large-scale structure. (Reproduced courtesy of NRAO.)

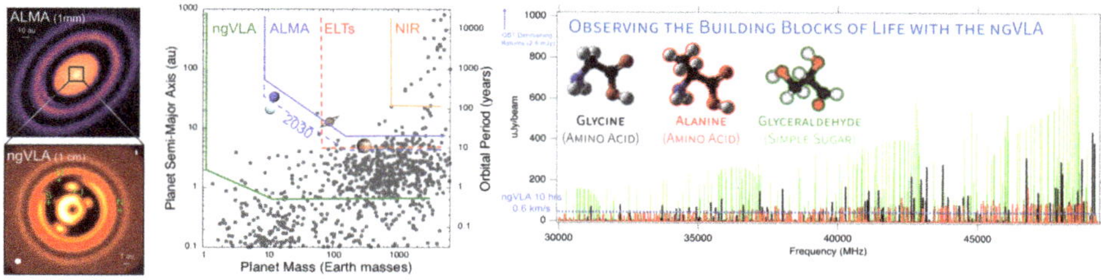

Figure 7.4. **Left (a):** A comparison of ngVLA capabilities with ALMA and other large telescopes. Upper left inset shows the star HD163296 as imaged by ALMA at 1 mm with a resolution of 0.04″. This shows rings and gaps suggesting the presence of Saturn-size planets (Liu et al. 2018 and Zhang et al. 2018). The lower left inset shows a simulation of what ngVLA will see at 1 cm wavelengths of the innermost region, if the star has Jupiter, Saturn, Uranus, and Neptune-size planets. The right panel shows the detection limits of ngVLA, current ALMA, projected 2030 ALMA, and other telescopes. (Reproduced courtesy of ALMA.) **Right (b):** The ngVLA high-sensitivity and spectral frequency coverage from 20 to 50 GHz gives it the capability to detect complex biogenic molecules that are predicted based on our understanding of chemical evolution. The simulation shows ngVLA's sensitivity for lines formed at temperatures of 80 K, relative velocity of 3 km s^{-1} in a 5″ source. (Reproduced courtesy of ALMA.)

7.2.2 Infrared

Many objects in the universe emit radiation that is too cool to be detected in visible light. There are some galaxies that are dominant in infrared, many stars that are much cooler than our Sun, and interstellar molecules. Infrared radiation can also penetrate dusty regions of space where optical light is easily scattered, allowing us to see deeper into the center of our Galaxy and into dusty, gas-filled star-forming regions. Our own Earth is a factor of a billion fainter than our Sun in optical wavelengths but only a factor of a million fainter in infrared wavelengths: this is because it absorbs the visible light from our Sun and reradiates it in infrared. This means that Earth-like planets associated with Sun-like stars are 1000 times easier to detect in infrared radiation than in the optical (although the shorter wavelength of visible light provide some competing advantages, and both wavelength bands are under study for detection of distant Earth-like planets).

The effects of the Big Bang also mean that gathering information from the farthest regions of the universe requires the infrared. As we learned in Chapter 6 (From Energy to Matter) the universe is expanding and most of the galaxies within it are moving away from each other; and the farther the Galaxy the faster it is moving. When an object is moving away from us, the light that it emits is redshifted so the wavelengths get longer and thereby shifted from optical toward the infrared.

7.2.2.1 The James Webb Space Telescope (JWST)

Launched in December 2021, JWST is an international collaboration between NASA, the European Space Agency (ESA), and the Canadian Space Agency (CSA) designed to complement and extend the discoveries made with Hubble. It operates primarily in the infrared, overlapping the wavelengths covered by both HST and Spitzer but with a much greater collecting area (six times the area of HST and more than 50 times the area of Spitzer; Figure 7.5(a)). The large collecting area is possible because of new technologies developed to make lightweight deployable mirrors. Lightweight means

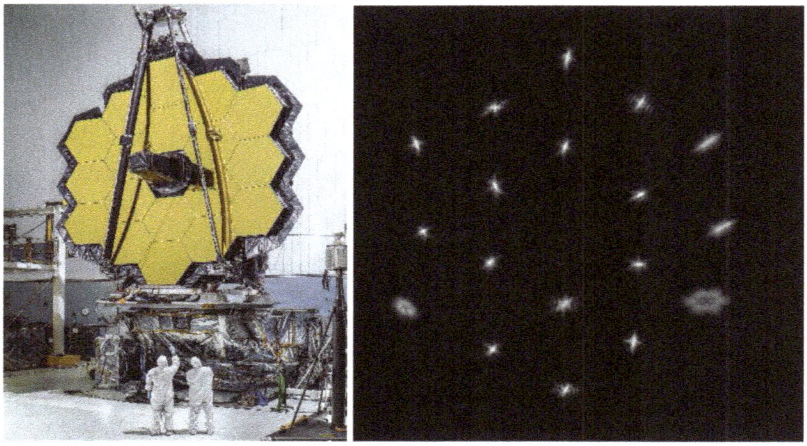

Figure 7.5. **Left (a):** JWST's 18 gold plated beryllium mirror segments form the equivalent of a 6.5 m diameter mirror. They have more than 6 times the collecting area of Hubble (2.4 m single mirror). JWST can detect objects up to 100 times fainter than Hubble, and that formed much earlier in the history of the universe. (Reproduced courtesy of NASA/KPL.) **Right (b):** An animation showing the 18 images of the star HD 84406 before and after focusing. (Reproduced courtesy of NASA.) Animation available at http://doi.org/10.1088/2514-3433/ac81b3.

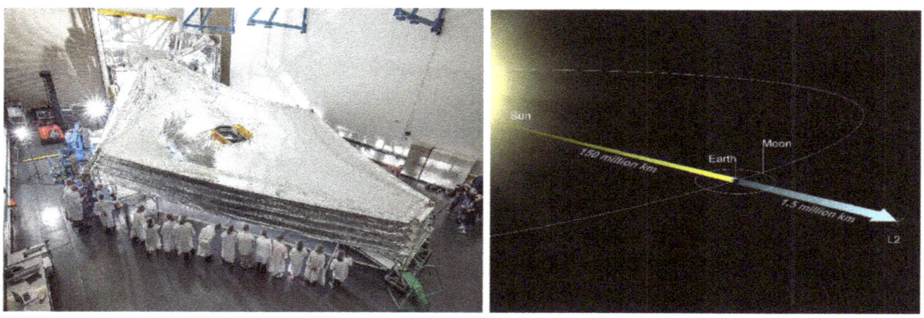

Figure 7.6. **Left (a):** The sunshields are five layers deep and essential for maintaining the low temperatures (less than 50K) required so that radiation from the telescope itself does not contaminate the observations. (Reproduced courtesy of NASA.) **Right (b):** JWST's orbit is located at the 2nd Lagrangian point of the Earth-Sun system, a position that allows it to keep the Sun, Moon, and Earth (all sources of infrared radiation) on the same side so that the sunshields are always protecting the telescope. (Reproduced courtesy of NASA.)

that in spite of the increase in size, JWST's mass is about half that of Hubble. Deployable means that the mirrors and sunshields are built to fold to fit into the restricted space of the launch vehicle and then unfolded when in space. This was a spectacular achievement. JWST is also placed at the 2nd Lagrangian point (see Chapter 5) so that the sunshields protect it from the heat of both the Earth and the Sun to maintain the cold operating temperatures needed for infrared detection (Figure 7.6). The placement also means that it can more efficiently observe the sky (Hubble's view of its targets is typically blocked by Earth for almost half its orbit).

JWST came into full focus in April of 2022 (Figure 7.7). In order to tune the four instruments on the spacecraft, they were pointed at the Large Magellanic Cloud.

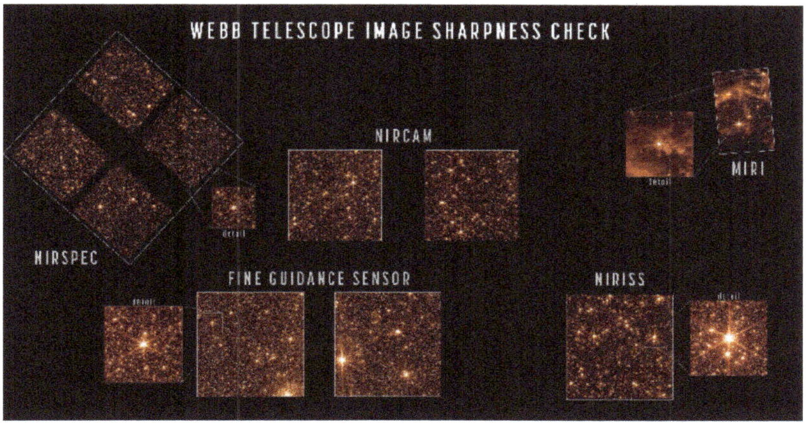

Figure 7.7. These JWST images of the Large Magellanic Cloud (a small satellite galaxy of the Milky Way) show the instruments in full focus in April of 2022. The sizes and positions of the images depict the relative arrangement of each instrument in the focal plane of the telescope, each points at a slightly offset part of the sky relative to one another. The three imaging instruments are NIRCam (images shown here at a wavelength of 2 microns), NIRISS (image shown here at 1.5 microns), and MIRI (shown at 7.7 microns, a longer wavelength revealing emission from interstellar clouds as well as starlight). NIRSpec is a spectrograph rather than imager but can take images, such as the 1.1 micron image shown here, for calibrations and target acquisition. The dark regions visible in parts of the NIRSpec data are due to structures of its microshutter array, which has several hundred thousand controllable shutters that can be opened or shut to select which light is sent into the spectrograph. Lastly, the Fine Guidance Sensor tracks guide stars to point the observatory accurately and precisely; its sensors are not generally used for scientific imaging but can take calibration images such as those shown here. This image data is used not just to assess image sharpness but also to precisely measure and calibrate subtle image distortions and alignments between sensors as part of the overall instrument calibration process. (Reproduced courtesy of NASA.)

This provided a field of view of hundreds of thousands of stars. JWST's greater collecting area allow it to see objects that are at far greater distances or are factors of 10 to 100 fainter than that visible to Hubble (Figure 7.5(b)): it will detect the first stars forming and combining to form conglomerates of stars that then collide and merge to form ever bigger galaxies and ultimately clusters of galaxies. Finding these first sources and watching the process from when the universe was only a few hundred million years old to now is critical to understanding the formation of the structure (from the smallest to the largest scale) of the universe.

7.2.2.2 The Nancy Roman Space Telescope

The 2010 decadal survey[1] gave the highest priority in space missions to the Wide Field InfraRed Survey Telescope (WFIRST), now renamed the Nancy Roman Space

[1] Since 1950 astronomers and astrophysicists in the US have undertaken decadal surveys (sponsored by the National Research Council) that draw on input from the entire astronomical community to determine a consensus opinion on the most effective way to utilize available funding to increase instrumental and research infrastructure. The 2010 decadal survey (New Worlds, New Horizons in Astronomy and Astrophysics) chaired by Roger Blandford, and the 2020 decadal survey (Pathways to Discovery in Astronomy and Astrophysics for the 2020s), chaired by Fiona Harrison and Robert Kennicutt lay out ambitious plans for sustaining the current progress in research.

Telescope, it is due to be launched in the mid-2020s. Roman's Wide Field Instrument (WFI) consists of a 300 megapixel infrared camera that has a field-of-view 100 times that of Hubble (Figure 7.8). This makes it ideal for surveys—something that JWST, designed for pointed observations, will not be able to do. The WFI will be able to look back to 150 million years after the Big Bang; significantly farther back than HST which can look back to 400 million years after the Big Bang. Roman will survey thousands of degrees in the sky at high speeds with the same sensitivity and resolution as HST in the optical.

To search for exoplanets, Roman will monitor the central bulge of the Milky Way for small deviations in brightness due to microlensing by intervening stars and their associated planetary systems (Figure 7.9). Roman's census of microlensed planets will complement planets found by the Kepler mission (Chapter 3 To Dust; Figure 7.10(a)). While Kepler's transit method is optimal for finding planets that are close to their parent star, Roman's microlensing technique will allow it to detect planets that are far from the parent stars, as well as those that may not be bound to any star. Together they will determine how common Earth-like planets are over a wide range of orbital parameters.

The Coronagraph Instrument (CI) on Roman, currently under construction by the Jet Propulsion Laboratory, will enable direct imaging of exoplanets by blocking the light of the star they orbit. Two major technology milestones make this the most powerful coronagraph to be flown to date: (1) flexible mirrors whose shape can be changed in real time by thousands of actuators in order to correct small changes in the optics, (2) special masks that compensate for the bending of light around the edges around light-blocking elements. The coronagraph will allow Roman to detect

Figure 7.8. The Nancy Roman space telescope with a 2.4 m diameter mirror is due to be launched in the mid-2020s. It will provide near-infrared imaging and low-resolution-spectroscopy. Roman's wide-field-of-view (100 times that of Hubble) at the same sensitivity and resolution of Hubble makes it ideal for surveys. The animation compares the two FOVs. (Reproduced courtesy of NASA.) Animation available at http://doi.org/10.1088/2514-3433/ac81b3.

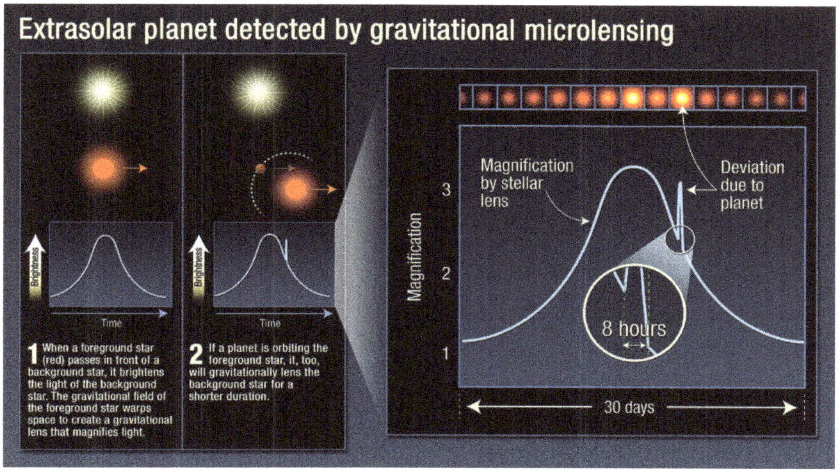

Figure 7.9. Microlensing is a type of gravitational lensing but with much smaller lenses: stars and planets rather than galaxies and clusters of galaxies. If an observer sees a distant source star pass behind a nearby star and associated planet, the distant source will be brightened twice: once by the star and once by the planet, with lensing by the planet occurring on a much shorter timescale. (Reproduced courtesy of NASA/JPL/ESA/Sahu.)

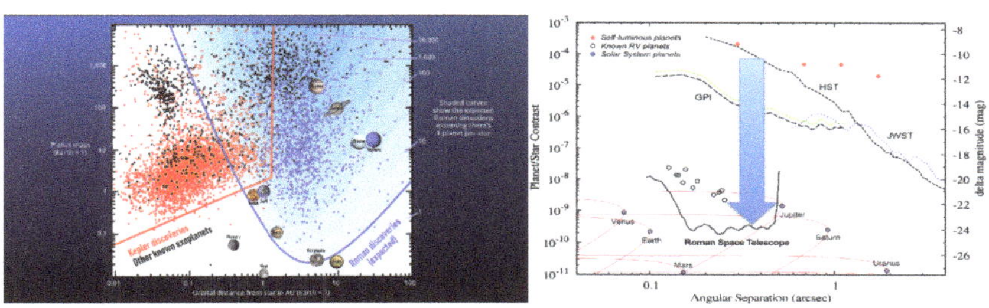

Figure 7.10. Left (a): Exoplanet discovery space for Roman using microlensing. (Reproduced courtesy of NASA.) **Right (b):** The detection sensitivity of the Roman Coronagraph Instrument compared to HST and JWST. (Reproduced courtesy of NASA.)

planets that are almost a billion times fainter than their host star (Figure 7.10(b)). In addition to detecting faint exoplanets, the coronagraph will also obtain clear images of the disks of stars providing insight on early stages of the star's life.

The cause of the accelerated expansion of the universe is one of the most important and profound scientific questions of our time. The currently postulated cause of the accelerating expansion of the universe is dark energy. Roman's wide field-of-view will allow it to form a census of galaxies in order to study their expansion history statistically. Understanding the growth rate of large-scale structure in the universe will have direct bearing on our understanding of the nature of dark energy.

7.2.3 Optical

Adaptive optics was suggested as early as 1953 by Horace Babcock (Babcock 1953) but the computer technology necessary to utilize it did not become available until the

1990s. Although optical light easily penetrates the Earth's atmosphere it is distorted by turbulence in the atmosphere. Before adaptive optics the resolution of a ground-based telescope was limited by this turbulence to roughly 1 arcsec (arcsecs are measures of angular size: 1 arcsec is 1/360th of a degree; it is roughly the size a penny would appear at a distance of 3.2 km (2 miles)). Although the Hubble Space Telescope's location in orbit eliminated atmospheric turbulence, it could provide only the resolution allowed by its 2.4 m mirror. Today, with adaptive optics, several ground-based telescopes can match or exceed Hubble's resolution and, in planning, are large telescopes that will far exceed it.

The application of adaptive optics to optical telescopes resulted in something many astronomers had not expected to see in their lifetimes: the image of a planet from outside our own solar system (see Chapter 3, Figures 3.15 and 3.16). All wavelengths are also benefiting by using telescopes with large fields of view that allow surveys of the entire sky. Clearly, the more of the sky you cover, and the oftener you look, the more likely you are to find something unusual.

7.2.3.1 The Vera C. Rubin Telescope

The 2010 decadal survey's highest rated ground-based facility, the Large Synoptic Survey Telescope (LSST), now named the Vera C. Rubin Telescope, continues as a priority item with the 2020 survey. The Rubin telescope will pair the world's largest digital camera (3200 Megapixels), with one of the world's largest mirrors (8.5 m or 27.5 ft in diameter; Figure 7.11(b)), with a wide field-of-view enabling it to produce images of the entire available sky every 3 nights. That translates to 30 terabytes of data daily. Rubin's ability to detect faint objects in 10–20 s exposures makes it ideal for detection of near Earth objects (NEOs), which move rapidly across the field of view or faint transients that only exist for seconds. The sequential images of the entire observable sky can either be combined to provide an image of unprecedented depth, subtracted from each other to highlight objects that change in brightness or position, or be concatenated to produce movies of objects that change or move. It's location in the sparsely settled and dry climate of Cerro Pachon in Chile will result in superb deep images of the greater part of the Milky Way from the ultraviolet to

Figure 7.11. Left (a): The Vera C. Rubin summit facility design takes into consideration the winds and topography of the its location in the dry climate of Cerro Pachon in Chile. (Reproduced courtesy of Rubin Obs/NSF/AURA.) **Right (b):** It took two days to load 51,900 pounds of glass into the furnace molding for the LSST primary mirror. (Reproduced courtesy of Rubin Obs/NSF/AURA.)

near-infrared. Compared to the best currently available optical survey (Sloan Digital Sky Survey; SDSS; see Chapter 6, Figures 6.9 and 6.10), Rubin will cover more than twice the area, with hundreds of observations of each region (instead of just one or two), with each individual observation detecting objects fainter by a factor of six. The coverage of the plane of our Galaxy will yield data for numerous star-forming regions, and its infrared capability will allow it to penetrate the interstellar dust, detecting on the order of 10 billion stars.

7.2.3.2 Giant Segmented Mirror Telescopes (GSMTs)

While the Rubin will be able to survey the sky efficiently, its ability to see distant faint sources is limited by its size. The GSMTs will combine interferometry techniques with adaptive optics. The combined increase in light gathering power and angular resolution will make optical images 10–12 times sharper than Hubble's and match the resolution ALMA achieves at radio wavelengths. The GSMTs under development will be able to extract spectral information from the atmospheres of planets, enabling searches for molecular signatures of life. These telescopes will enable the use of tomographic techniques to resolve X-ray binaries in external galaxies.

One concept for a GSMT is the Giant Magellan Telescope (GMT), a collaboration of ten institutions from three countries (US, Australia, and Korea). Composed of 7 mirrors, each the size of the Rubin mirror (8.5 m); GMT's combined angular resolution will match that of a 25 m (80 ft) mirror (Figure 7.12(a)). Another concept is the Thirty Meter Telescope (TMT), an international partnership of six institutions from five countries (US, Canada, Japan, China, and India). TMT's mirror will be made up of 492 individual segments. Precisely aligned, these segments will work as a single reflective surface 30 m across. Both telescopes have identified high altitude sites that provide the dark skies, dry, and stable atmospheric conditions ideal for astronomy (Figure 7.12(b)). Official construction of the GMT began in 2015 with a groundbreaking ceremony at Las Campanas, Chile. Ideally, astronomers would like such facilities in both hemispheres in order to study the whole sky.

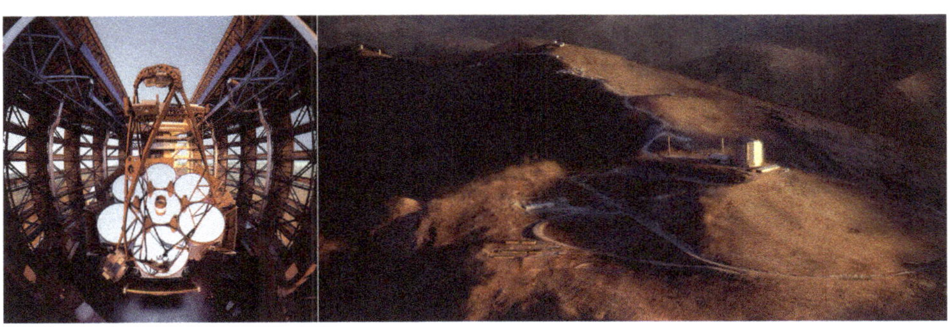

Figure 7.12. Left (a): The seven mirrors that comprise the GMT. (Reproduced courtesy of Giant Magellan Telescope—GMTO Corporation). **Right (b):** Foreground building is an artists concept of the Giant Magellan Telescope at the Las Campanas Observatory (LCO) in Chile. (Reproduced courtesy of Giant Magellan Telescope—GMTO Corporation).)

7.2.4 X-Ray

While "visible" light from the early galaxies must be studied in the infrared, X-rays can penetrate the absorbing gas and dust to study the supermassive black holes at the centers of those early galaxies. The bulk of radiation emitting matter in our universe is comprised of very hot gas which fills the space between galaxies and clusters of galaxies and can be studied only with space borne X-ray instruments. X-rays also allow us to study the most important energy generating process in the universe such as accretion onto black holes.

Perhaps one of the last remaining frontiers to be explored is polarimetry from X-rays. Polarization of electromagnetic waves conventionally refers to the direction of the electric field (see Figure 1.2). Studying polarization gives important information about the magnetic fields and the geometry of sources that cannot be resolved spatially. X-ray sources generally have strong magnetic fields, show synchrotron emission, and undergo Compton scattering. One of the first detections of X-ray polarization was reported from the Crab in 1976 by Martin Weisskopf and collaborators (Weisskopf et al. 1976). The instrument used was launched by NASA on the 8th Orbiting Solar Observatory (OSO-8) in 1975. Only upper limits were obtained for a variety of other bright X-ray sources. In 2017, the Cadmium Zinc Telluride Imager (CZTI) on India's AstroSat observatory (launched in 2015), demonstrated that the amount of polarization from the Crab pulsar varies strongly with the spin of the pulsar (Vadawale et al. 2018; Figure 7.13).

7.2.4.1 Imaging X-Ray Polarimetry Explorer (IXPE)

Successfully launched in December of 2021, IXPE has two orders of magnitude increase in sensitivity over the instrument on OSO-8 (Figure 7.14(a)); Weisskopf et al. 2016). A collaboration between NASA's Marshall Space Flight Center and the Italian Space Agency (ASI), IXPE's polarization measurements are made possible by new detector technology contributed by ASI. Marshall's responsibility was to build and calibrate the mirror assemblies and deliver them to Ball for assembly,

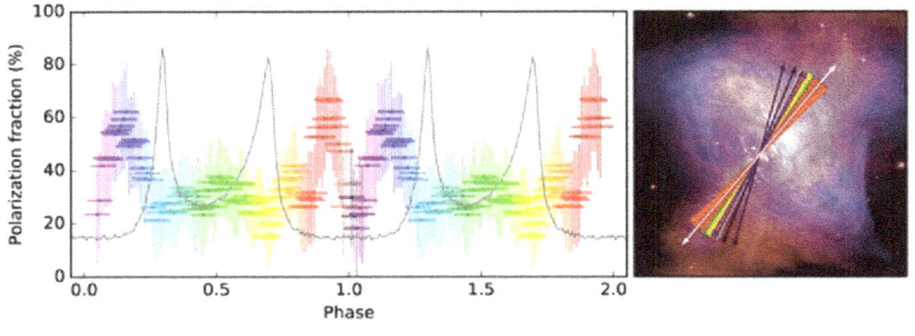

Figure 7.13. Left (a): An image showing two cycles of the pulse profile of the Crab pulsar. The colored bars shows the amount of polarized radiation as a function of phase as measured by the CZTI instrument on India's AstroSat. (Reproduced courtesy of NASA/CXC/ASU.) **Right (b):** The polarization depicted in the left panel superposed on a composite X-ray (Chandra Observatory) and optical (Hubble) image of the Crab Nebula relative to the pulsar's jet (depicted in white). Colors represent the phase bin as shown on the left. (Reproduced courtesy of NASA/CXC/ASU.)

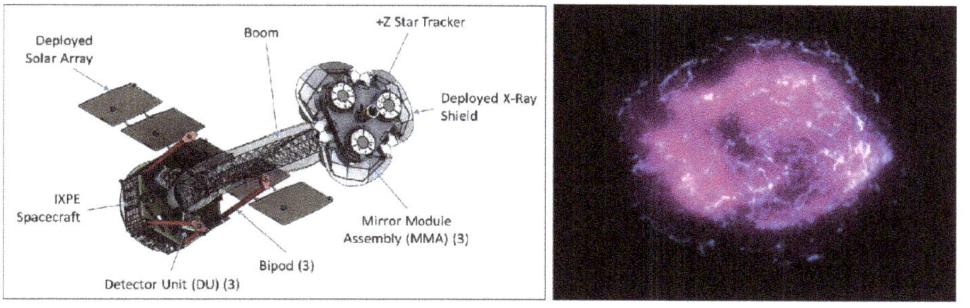

Figure 7.14. **Left (a):** Launched in December of 2021, NASA's Imaging X-ray Polarimetry Explorer (IXPE), contains three identical X-ray telescopes, consisting of 24 nested cylindrical mirrors paired with polarization-sensitive detectors. (Reproduced courtesy of NASA.) **Right (b):** An image of the supernova remnant Cassiopeia A combining the first X-ray data from IXPE (Magenta) with observations from the Chandra X-ray Observatory (Blue). (Reproduced courtesy of NASA.)

integration, and environmental testing. Within a few months of its launch, IXPE, observed polarization from three of its first six targets. In addition to confirming polarization from the Crab, IXPE has also detected polarization from Hercules X-1, one of the first confirmed pulsars, and the magnetar 4U 0142+61, in the constellation of Cassiopeia (Figure 7.14(b)). Analysis of results is still underway. Figure 7.15 shows model expectations for polarization from Hercules X-1 (Caiazzo & Heyl 2021a, 2021b). IXPE should provide information on the magnetic field structure of the X-ray emitting regions of systems under extreme physical

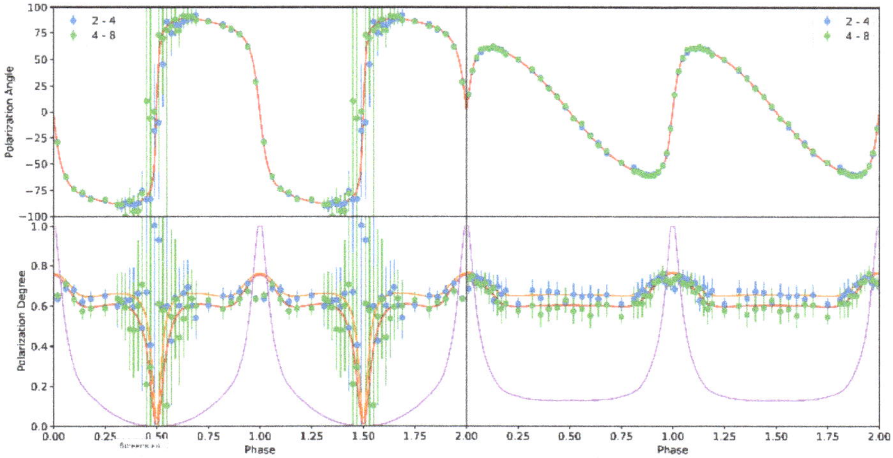

Figure 7.15. Predicted polarization of Her X-1 with a 70 ks observation with IXPE. The purple curves in the lower panels depict the total counts from the accretion column in the model. The left panels show the result with a single accretion column and the right panels depict two accretion columns. The large error bars near phase 0.5 in the left panel result from the lack of flux from the accretion column at this phase. The slight discrepancy near phase 1.0 between the measured and model polarization fraction results from smearing the polarization direction over the bin and can be avoided with an unbinned analysis. (Reproduced courtesy of Jeremy Heyl (2022).)

conditions: neutron stars, black holes, supernova remnants, and AGN. The addition of a dedicated instrument to a new method of seeing, in this case polarization, may also lead to detection of new classes of objects.

7.2.4.2 X-Ray Imaging and Spectroscopic Mission (XRISM)

Due to be launched in 2023, XRISM, led by the Japan Aerospace Exploration Agency (JAXA) with contributions from NASA and ESA is meant to provide continuity in soft energy X-ray capability through the 2020s. The current state-of-the-art workhorses, the Chandra X-ray Observatory and the XMM-Newton Observatory are both over 20 years old with detectors gradually aging. The next generation X-ray telescope, Athena, is not due to be launched until the 2030s. The satellite XRISM will fly two soft X-ray telescopes (Figure 7.16) one paired with a X-ray Calorimeter Spectrometer providing high-energy resolution (RESOLVE) and one with a wide-field CCD detector (Xtend). While XRISM does not have the high angular resolution of Chandra (1″) or XMM (15″), the wide-field (30′ by 30′) of Xtend makes it particularly well suited to study the large-scale structures of the universe, and the energy resolution of the calorimeter will enable unprecedented high resolution X-ray spectroscopy. XRISM will serve to test technology that will be used for later generations of X-ray satellites.

7.2.4.3 Advanced Telescope for High-ENergy Astrophysics (Athena)

Approved by European Space Agency (ESA) in 2014, Athena has an expected launch date in the 2030s. It is the major X-ray telescope in planning over the next 10–20 years and will maintain a window on high-energy phenomena in many astrophysical contexts, from the solar system to the most distant objects known.

Athena has a single telescope with a focal length of 12 m which combines high collecting area with low mass (Figure 7.17(a)). It will exceed the collecting areas of Chandra and XMM by at least one order of magnitude (Figure 7.17(b)). The angular resolution of 5″ is less than Chandra (1″), but a factor of three better than XMM (15″), more than sufficient to distinguish AGN from other background sources.

There are two focal plane instruments: the X-ray Integral Field Unit (X-IFU) and the Wide Field Imager (WFI) providing breakthrough capabilities. The combination

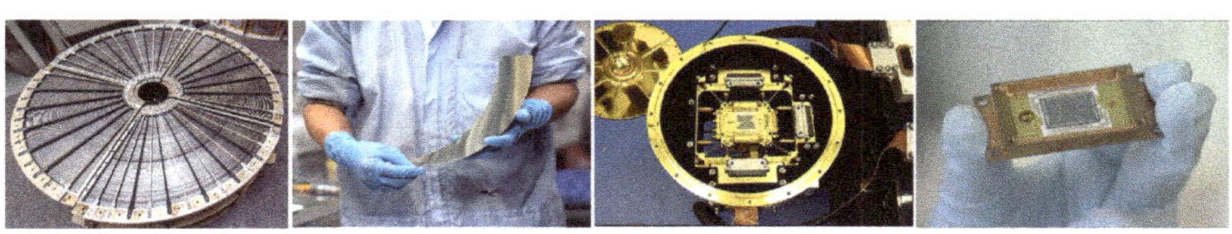

Figure 7.16. Left (a): One of the two X-ray Mirror Assemblies to be flown on XRISM. Composed of four quadrants each with 203 nested mirror segments. Primary segments are parabolic and secondary segments are hyperbolic providing spatial resolution of 1.7 arcmin. (Reproduced courtesy of NASA/HEASARC.) **Center left (b):** One of the 1624 foil segments that constitute the XMA. (Reproduced courtesy of NASA/HEASARC.) **Center right (a):** The micro-calorimeter to be flown on XRISM provides 5–7 eV energy resolution in the 0.3–13 keV band with a 3 × 3 arcmin field-of-view. (Reproduced courtesy of NASA/HEASARC.) **Right (b):** Close-up showing the size of the micro-calorimeter which is a 6 × 6 grid of 30″ pixels. It needs to be kept at a temperature of 50 mK. (Reproduced courtesy of NASA/HEASARC.)

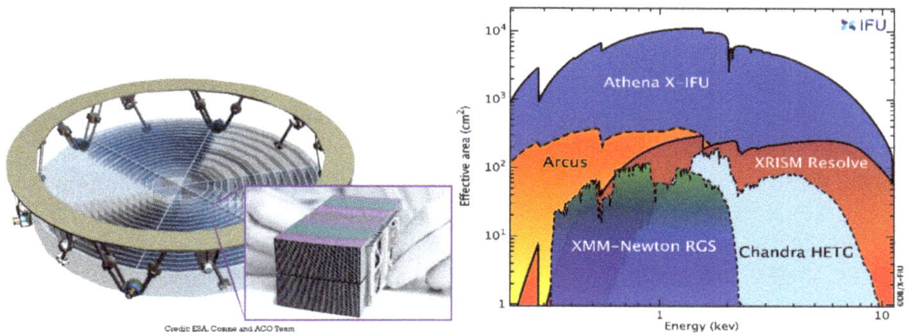

Figure 7.17. Left (a): Schematic design of the mirror assembly for Athena: inset shows mirror modules and stacks. (Reproduced courtesy of ESA, Cosine and the ACO Team.) **Right (b):** The X-IFU effective area compared to other imaging (XRISM) and dispersive (XMM-RGS, Chandra HETG, Arcus) spectrometers. Data provided by Randall Smith for the Arcus effective area. (Reproduced courtesy of X-IFU consortium.)

of wide-field images and high resolution spectroscopy in X-rays will enable Athena to study how large-scale hot gas structures form and evolve. The X-IFU consortium has participation from 11 ESA member states, as well as Japan (JAXA), and the United States (NASA). Covering the 0.3–10 keV energy band, it will consist of thousands of micro-calorimeters providing unprecedented energy resolution (2.5 eV) over a large field of view (5 arcmin diameter). The X-IFU will obtain images of astrophysical objects where every pixel of the image will produce a full spectrum. This detail will enable the study of how SMBHs interact with galaxies, and to measure abundances of different elements in galaxy clusters. The WFI consortium is led by the Max Planck Institute for extraterrestrial Physics in Germany along with partners from 11 countries. The WFI covers the 0.2–15 keV energy range with a large field of view (40′ by 40′), excellent angular resolution (5″) and simultaneous high-time resolution and high count rate capabilities. As a survey instrument it will surpass all currently existing capabilities. WFI will enable Athena to undertake a complete census of SMBHs, including ones that are heavily obscured or at high redshift.

7.2.5 Gamma-Ray Instruments

While space facilities such as the Compton and Fermi Gamma-Ray Observatories have given us amazing views of the gamma-ray universe, the very high energy (VHE) range needs to be studied from the ground. Although extremely energetic, VHE gamma-ray photons are few in number and the collecting areas needed to study them are too large to be flown.

Fortunately, a technique first developed by astronomers at the Whipple observatory in Arizona allows efficient study of VHE gamma-rays from the ground. As VHE gamma-rays come into the upper atmosphere, they create cascades of subatomic particles. These particles contain the tremendous amount of energy which the gamma-ray had originally, and consequently, they travel at very high speeds—in fact faster than the speed of light in air. When a particle goes faster than the speed of light in the medium in which it travels (e.g., water or air) it produces the electromagnetic equivalent of a sonic boom (which is produced by something traveling faster than the speed of sound in air). This electromagnetic boom is called Cherenkov Radiation and consists of a brief flash of light (See Chapters 1 and 4). This light can be detected by using very big mirrors and highly sensitive, fast light detectors called photomultiplier tubes (PMTs).

Such ground-based imaging Cherenkov telescopes (ICTs) contribute to both astrophysics and particle physics. The Whipple observatory used a 10 m ICT to discover the first known source of TeV gamma-rays (the Crab Nebula) in 1989 (Weekes 1989). Since then nearly a hundred TeV-sources have been detected by ground-based gamma-ray telescopes such as MAGIC, the High Energy Stereoscopic System (H.E.S.S.) (Figure 7.18(a)), and CANGAROO. Many of the TeV-sources are associated with known objects such as supernova or pulsars, but there are also mystery sources that have not been seen at any other wavelengths, but may be connected with galactic cosmic rays. VERITAS (Very Energetic Radiation Imaging Telescope Array System) replaced the Whipple 10 m ICT with four 12 m ICTs and serves as an effective complement to NASA's Fermi Space Observatory.

7.2.5.1 Cherenkov Telescope Array (CTA)

The design of CTA is based on currently available technology and incorporates three mirror sizes to encompass the full energy range available from the ground (tens of GV to about 300 TeV; Figure 7.18(b)). CTA is aimed to increase sensitivity by a factor of 10 in the current energy domain of about 100 GeV to some 10 TeV and to extend the accessible energy range well below 100 GeV and to above 100 TeV.

The telescopes will be placed on two sites with slightly different science goals. The Southern hemisphere site, located near ESO's Paranal Observatory in Chile will have all three mirror sizes: four 23 m diameter dishes, twenty-five 12 m diameter dishes, and seventy 4 m diameter dishes. The Northern Hemisphere site is near the Instituto de Astrofisica de Canarias, Roque de los Muchachose Observatory in La Palma (the site of the Major Atmospheric Gamma Ray Imaging Cherenkov (MAGIC) telescopes) and will have four 23 m size and fifteen 12 m size dishes.

Figure 7.18. Left (a): The High Energy Stereoscopic System (H.E.S.S.) located in Namibia consists of four 12 m diameter imaging atmospheric Cherenkov telescope, and one IACT of 28 m diameter. (Reproduced courtesy of Klepser, DESY, H.E.S.S. collaboration.) **Right (b):** The Cherenkov Telescope Array, to be located near ESO's Paranal Observatory in Chile, will have three sizes of mirrors. For the highest energies (up to 300 TeV) 4 m size dishes, for the lowest energies (down to 10s of GeV) 23 m size dishes, and 12 m size dishes for energies between 100 GeV and 1 TeV. (Reproduced courtesy of CTA/M-A. Besel/IAC (G.P. Diaz)/ESO.)

The La Palma site with the low-energy instrumentation (10s of GeV to about 1 TeV) will concentrate on distant extragalactic sources such as AGN. The Paranal site which covers the full energy range will concentrate on deep investigations of galactic sources and the central part of our Galaxy.

The CTA will have a large impact on astrophysics and fundamental physics research. It will contribute to the study of the origin of cosmic rays and their impact on the constituents of the universe, the investigation of the nature and variety of black hole particle accelerators, searching for dark matter and effects of quantum gravity, and the inquiry into the ultimate nature of matter and physics beyond the Standard Model.

7.2.6 Time Domain Astronomy/Multiwavelength Telescopes

Each of the efforts listed above necessarily relies on wavelength-dependent technology and therefore must concentrate on specific niches of the spectrum. A broad range of parameter space that has not been well explored is time variations covering multiple wavebands.

The vast majority of strange new objects and phenomena found by space borne instruments turn out to be variable on multiple timescales. For pulsars (whose variations timescales are measured in seconds or less), or even for active galactic nuclei (where variations can occur on the order of days) it is essential to have *simultaneous* observations at multiple wavelengths and channels. Indeed, simultaneous observations can yield fundamental information that is obtainable in no other way. However, the task of coordinating such observations with telescopes on the ground and in space are difficult and sometimes impossible.

The high angular resolution and state-of-the-art sensitivity of observatories like HST and JWST make them more valuable for a qualitatively different observing program. Simultaneous observations of *selected* objects that require their unique capabilities can be done, but the vanishingly small probability of scheduling long, repeated observations of many targets with different variability timescales makes it impossible to conduct systematic surveys of entire classes of objects.

As an example, consider coordinated observations using JWST for infrared, HST for ultraviolet, and a ground-based optical telescope. Ground-based telescopes are hampered by day/night boundaries incommensurate with the satellite-borne telescopes and by unpredictable weather conditions. Even if a target is accessible to two satellites at the same time, differing orbits can still prevent truly simultaneous observations. For satellites with short periods (96 min for HST) earth eclipses and South Atlantic Anomaly passages (areas where Earth's inner Van Allen radiation belt comes close to the Earth's surface, causing increased flux of energetic particles that is problematic for orbiting satellites) cause frequent interruptions.

These questions require an instrument complement with the greatest practical wavelength coverage, moderate spectral resolution, photon counting detectors with high-time resolution, well-established relative flux calibrations, and sufficient sensitivity to detect enough sources to form good statistical samples in a reasonable amount of time. LUVOIR is such an instrument.

7.2.6.1 Large Ultraviolet Optical Infrared Surveyor (LUVOIR)

The simultaneous, multifrequency observation of astrophysical objects in an unbiased fashion would provide an entirely new capability for astronomers, allowing new phenomena to be discovered.

Currently two significantly different design concepts have been proposed for LUVOIR (Figure 7.19(a)). The more ambitious one, LUVOIR-A, would combine 36 separate mirror segments into one that is 15 m in diameter: nearly forty times the collecting power of either HST or the Nancy Roman Space Telescope (Figure 7.19(b)) and twenty-four times their resolving power.

LUVOIR-A will have four instruments: the Extreme Coronagraph for Living Planetary Systems (ECLIPS) will provide imaging and spectroscopy in three broad bands from 200 to 850 nm. The High definition Imager (HDI) has a wide field-of-view (2 × 3 arcmin) in order to benefit from LUVOIR's enhanced resolving power. The LUVOIR Ultraviolet Multi Object Spectrograph (LUMOS) has two orders of magnitude higher efficiency than the STIS on Hubble, its multi-object spectrograph will observe hundreds of targets simultaneously. The ultraviolet spectro-polarimeter, POLLUX, led by the French Space Agency (CNES) and a collaboration of 10 European institutions would provide high resolution (R = 120,000).

The complement of instruments comprising NASA's LUVOIR in combination with its high resolving power and collecting power would provide a unique new tool for astronomy, one that would make possible a wide range of observational programs that could not be done through coordinated observations with any existing or planned instruments.

LUVOIR will reach much farther than Hubble or NRST, to times when galaxies are just being formed. It will place limits on galaxy formation and on the nature of dark matter. LUVOIR will resolve star formation in galaxies far beyond our Local Group. LUMOS will characterize hundreds of exoplanets. and is particularly suitable for studying planetary atmospheres and ideal for searching for molecules that hint of life in the atmospheres exoplanets (e.g., water, ozone, carbon dioxide, and methane).

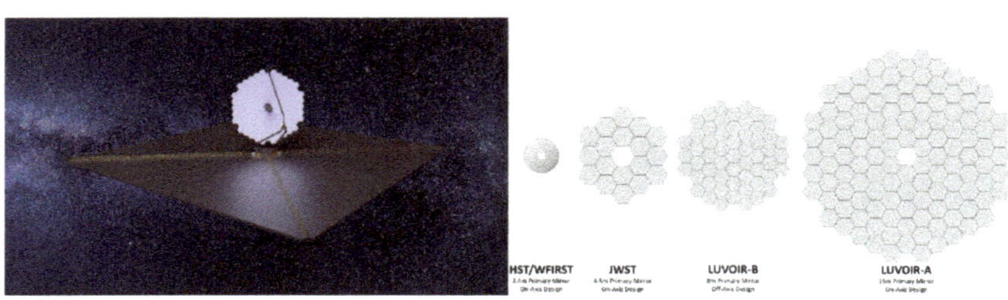

Figure 7.19. Left (a): An artist's concept of LUVOIR which will be placed at the Earth-Sun Lagrangian 2 point, enabling long points at individual sources. (Reproduced courtesy of NASA.) **Right (b):** LUVOIR will have up to a 15 m diameter primary telescope. More than twice the size of the JWST mirror (6.4 m diameter). (Reproduced courtesy of NASA.)

7.3 Cosmic Rays

The origin of Galactic and Ultra High Energy Cosmic Rays remains an open problem in astrophysics. Our understanding of CRs and UHECRs will have important input to the study of nucleosynthesis, dark matter, relativistic jets, particle acceleration and as yet unexplained transients. The Pierre Auger Observatory continues to be the state-of-the-art instrument for cosmic ray detection (see Chapters 1 and 6)

7.3.1 AugerPrime and AMIGA

The Pierre Auger Observatory found a sharp drop-off in cosmic rays above $10^{19.5}$ eV (see Chapter 1, Figure 1.16(a), with a suggestion that above the "knee" of the distribution ($10^{18.5}$ eV) one gets only particles more massive than protons (Meszaros et al. 2019). A major goal of AugerPrime is to understand what causes the flux suppression and generates the mass distribution of cosmic ray particles at different energies. It will also make a concerted search for protons at energies above the knee which will provide information on the flux of secondary gamma-rays and neutrinos. AugerPrime will add scintillator detectors (SSD) on top of each of the 1660 water-Cherenkov detectors (WCDs; see Chapter 1, Figure 1.17(b)); this means that shower particles will be sampled with two types of detectors which respond differently to muons and electromagnetic particles. This has the added advantage that it will be able to compare the SSD measurements directly with those of the Cherenkov Telescope Array experiment. AugerPrime will also replace the current electronics on each of the 1660 water-Cherenkov detectors with flexible surface detector electronics that allows for faster sampling, better timing accuracy, and increased dynamic range. The new electronics will serve to process both WCD and SSD signals.

AMIGA is a set of 61 muon counters spaced over 25 km^2, that will be buried 2.5 m underground in order to detect only the muon component of the air showers. This extends AugerPrime's ability to study lower energy cosmic rays. Pierre Auger also found that their observations of showers show inconsistencies with current models, suggesting a need for new hadronic interaction physics at energies beyond the reach of the LHC (Aab et al. 2016). By increasing the duty cycle for operations of the 27 fluorescence detectors (see Chapter 1, Figure 1.18) to 23%, they will trade higher sky background in order to follow shower maximum to energies beyond $10^{19.5}$ eV.

7.3.2 Dark Matter Particle Explorer (DAMPE)

Launched in December 2015 on a Long March 2D rocket, DAMPE is a collaboration of institutes from China, Italy, and Switzerland led by the Purple Mountain Observatory (PMO) of the Chinese Academy of Sciences (CAS). It was named Wukong after a Chinese classic hero whose name can be interpreted as Wu (understanding) and Kong (void or space). The Bismuth Germanium OXIDE calorimeter on DAMPE is the deepest calorimeter used in space. It is designed to detect high-energy cosmic rays at much higher energies (up to 100 TeV and at significantly higher energy resolution than existing space telescopes. The energy range and resolution gives access to heavy nuclei in the TeV energy range, as well as electrons, protons, and high-energy gamma-rays in order to study the origin, propagation, and acceleration mechanisms of cosmic rays and to search for evidence of dark matter (see Chapter 6). DAMPE measured the proton spectrum from 40 GeV to 100 TeV (An et al. 2019). The spectral hardening at a few hundred GeV is well measured by several detectors but the spectral softening at about 10 TeV is confidently detected by DAMPE. The spectral softening could be due to a cosmic ray source, such as a supernova remnant. A model fit to the DAMPE proton spectrum is consistent with the break energy due to the acceleration limit of the source. DAMPE also found a break in the spectrum of electron and positrons at 0.9 TeV (Ambrosi et al. 2017). This puts significant constraints on the candidates for dark matter particles.

7.4 Neutrinos

The detection of very high-energy neutrinos from outside the Galaxy by IceCube established the potential of neutrinos for multimessenger astrophysics and provided the incentive to upgrade the instrument and to extend its frequency range (Aartsen et al. 2019). Goals include identification of the sources of the highest energy neutrinos and cosmic rays, understanding the production mechanisms of high-energy cosmic particles, determining the contribution of active galaxies, stellar explosions, and hadronic emission from supernova remnants, and testing physics beyond the standard model.

7.4.1 IceCube-Upgrade

The IceCube-Upgrade, approved by the National Science Foundation in 2019 is expected to be constructed in the 2022–2023 polar field season. As described in Chapter 1 IceCube has three major components: the main array buried in a cubic kilometer of ice for detection of high-energy neutrinos (TeV range) consists of 86 boreholes drilled in the ice of Antarctica to depths of 2450 m; a surface cosmic ray air shower array (IceTop) to separate cosmic ray events from neutrino events; and a Deep Core where each borehole contains a string with 60 DOMs spaced five times closer together allowing for detection of neutrinos with energies below 100 GeV. IceCube-Upgrade will more than double the instruments to the Deep Core section reducing the spacing between strings and modules and the distance between light sensors by another factor of three (Figure 7.20). This will allow IceCube-Upgrade to reach energies as low as 10 GeV to study neutrino oscillations in the atmosphere, galactic supernova neutrinos, and other low-energy neutrino sources. The new instrumentation includes 750 advanced photodetectors and calibration devices which will apply to the full IceCube detector improving particle identification. The calibration devices enable a better understanding of the ice, and can be retroactively applied to archived events to improve on the resolution.

7.4.2 IceCube-Gen2

IceCube-Gen2 is possible because of the unexpectedly high clarity of the Antarctic ice found while constructing IceCube. This allows for addition of light sensors at twice the separation and covering 10 times the volume of the current array

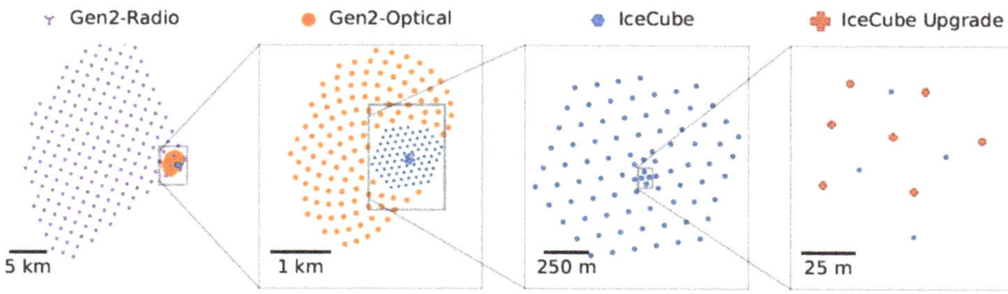

Figure 7.20. A schematic comparing IceCube (current), IceCube-Upgrade (2024), and IceCube-Gen2 (2033) Neutrino Observatory facility at the South Pole. Current IceCube shown with blue dots; IceCube-Upgrade shown as red dots; Gen2 will add a radio array consisting of 200 stations (purple dots); an optical high-energy array with 120 new strings (orange dots); IceCube-Gen2 strings are spaced 240 m apart and have optical modules over a vertical length of 1.25 km. The total instrumented volume in this design is 7.9 times larger than the current IceCube detector array. (Reproduced courtesy of the IceCube Collaboration.)

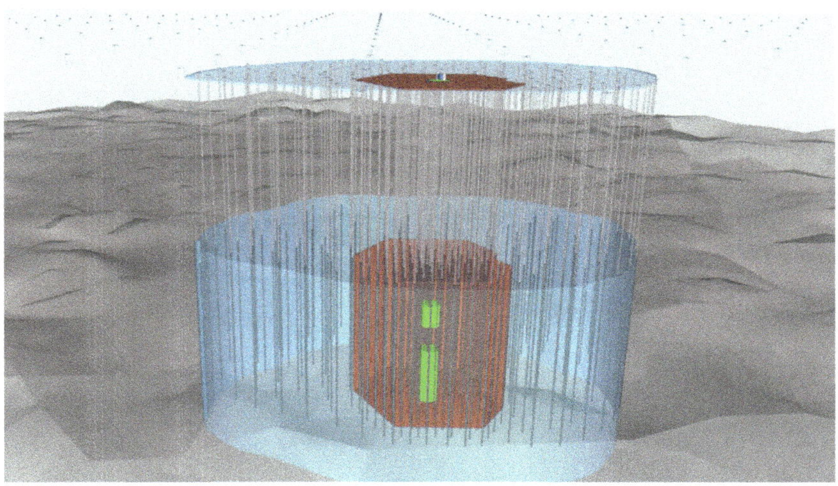

Figure 7.21. A schematic depicting the original IceCube and IceTop in red, the Deep Core in green; the increased area of IceCube-Gen2 in blue and a part of the 500 square kilometer radio array in gray dots. (Reproduced courtesy of the IceCube Collaboration.)

(Figure 7.21), leading to an order of magnitude improvement in neutrino detection rates. The 120 additional strings, spaced 240 m apart, will hold High Quantum Efficiency optical modules that collect three times the photons of the existing DOMs in the main array. This will allow the search for neutrinos with energies up to 10^{18} eV (EeV), probing fundamental physics with the highest energy particles in the universe. In addition, Gen2 will have a new surface component of radio detector stations covering nearly 500 km^2. The detection of radio emission will allow reconstruction of the energy of the shower and arrival direction of the neutrinos.

7.5 Gravitational Waves

Since the first detection of a direct gravitation wave in 2015 (Abbott et al. 2016), twenty-two gravitational wave events have been detected. Of these seventeen are binary black hole mergers (BBHs), two binary neutron star mergers (BNS), and three black hole–neutron star mergers. Of the 37 black holes involved in the mergers, 22 have masses that are much larger than have been measured for stellar mass black holes by other means. So one open question is why have high mass non-merging BHs not been detected? Also of interest is that all the LIGO/Virgo detected BHs have very low spins (spins that are not aligned with the orbital angular momentum) whereas non-merging BHs measured to date have mostly very large spins. Finally, of the twenty-two mergers only one has been associated with an electromagnetic counterpart. Clearly many more examples are needed with concerted effort to identify counterparts.

The gravitational wave spectrum covers over 18 orders of magnitude in frequency (Figure 7.22). Of these the LIGO/Virgo/Karga and other ground-based observatories currently under construction cover the highest frequencies. These frequencies allow the detection of compact binary mergers, and as they increase in sensitivity

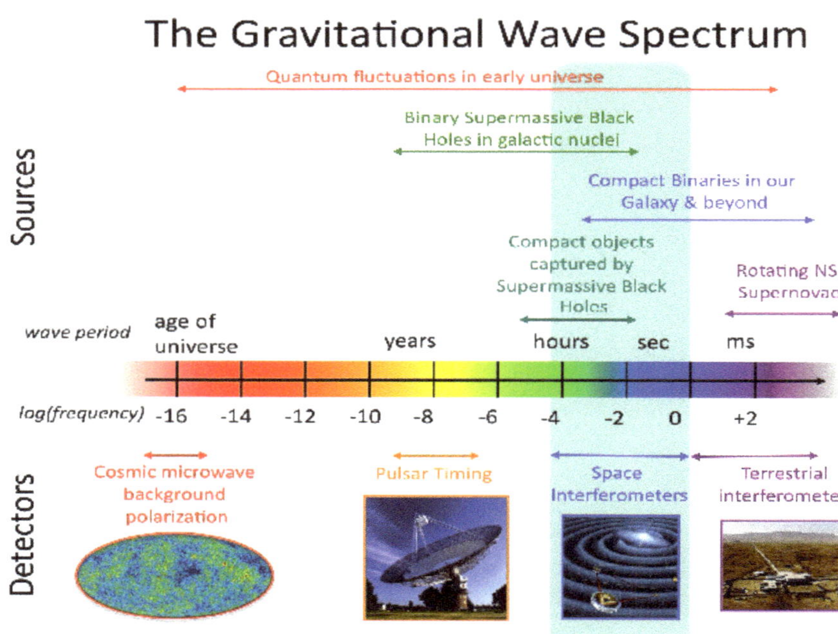

Figure 7.22. Gravitational waves cover over 18 orders of magnitude in frequency. Ground-based LIGO reaches periods of seconds to millisecond frequencies ($1–10^3$). Space based LISA (due for launch in the 2030s) will cover lower frequencies with periods of order hours to seconds ($10^{-3}–1$). Indirect techniques such as pulsar timing covers years to hours with frequencies from 10^{-9} to 10^{-3}, and CMB polarization covers events lasting from years to the age of the universe, back to the Big Bang: $10^{-16}–10^{-14}$. (Reproduced courtesy of NASA/Goddard Space Flight Center.)

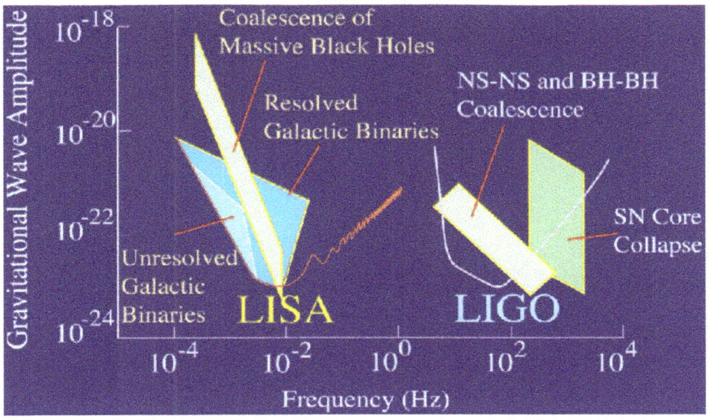

Figure 7.23. The range of gravitational wave amplitudes and frequencies that LISA will be and LIGO is sensitive to. Shaded areas show the frequency and amplitude of gravitational waves emitted by different types of objects (NS = neutron star; BH = black hole; SN = supernova). (Reproduced courtesy of NASA.

should be able to detect gravitational waves from rotating NS and supernovae. But in order to measure gravitational waves from SMBHs we need to reach much lower frequencies (Figure 7.23). This is the aim of the Laser Interferometer Space Antenna.

7.5.1 The "evolved" Laser Interferometer Space Antenna (eLISA)

eLISA, the evolved Laser Interferometer Space Antenna, is an ESA-led mission to observe astrophysical and cosmological sources of gravitational waves of low frequencies (0.1–100.0 mHz). This frequency band complements the frequencies studied by eLISA's earth-based counterpart LIGO (Figure 7.23), which is sensitive to higher frequencies.

LIGO is sensitive to gravity waves due to transient phenomena, like supernovae and the final minutes of the spiraling in of neutron star binaries, whereas LISA's frequency range is sensitive to the emission from mergers of massive black holes with masses ranging from 10,000 to 10 million solar masses; binaries that form after galactic mergers; compact stellar remnants as they slowly spiral to their final fate in the black holes at the centers of galaxies; millions of compact binaries in our own Galaxy; and possibly the waves generated shortly after the Big Bang.

eLISA will consist of three identical spacecraft flying in a triangular configuration, with equal arms of 5 million kilometers each (Figure 7.24). Each spacecraft will enclose a free-falling test mass.[2] As gravitational waves from distant sources reach eLISA, they warp spacetime, stretching and compressing the triangle in response to the "ripples" of spacetime caused by motion of the densest objects in the

[2] In December 2015, ESA launched the LISA pathfinder. The pathfinder used the same type of test masses that eLISA will use but they were separated by a mere 39cm encased within a single spacecraft. LISA Pathfinder was able to prove the feasibility of LISA.

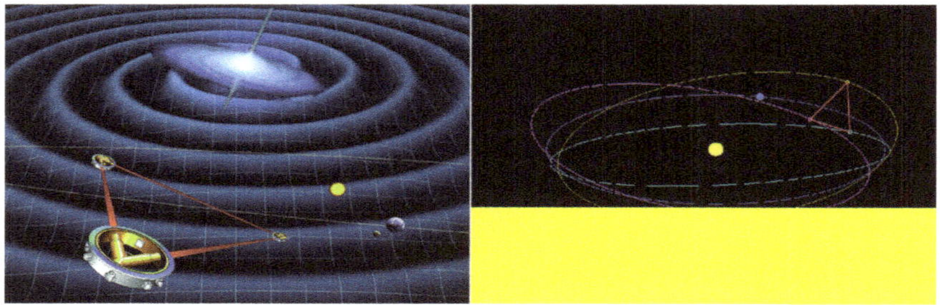

Figure 7.24. Left (a): eLISA will detect gravitational waves by accurately monitoring the separation between three spacecraft placed in a triangular configuration separated by over 5 million kilometers each. (Reproduced courtesy of NASA.) **Right (b):** An animation showing LISA's orbit with respect to our Sun and Earth. (Reproduced courtesy of ESA.) Animation available at http://doi.org/10.1088/2514-3433/ac81b3.

universe. The laser-connected spacecraft would measure their separations to an accuracy enabling detection of tens of pico-meters relative motions induced by passing gravitational waves. Precise monitoring of the separation between the spacecraft gives a measure of the size of the waves; the shape and timing of the waves provides information on the nature and evolution of the systems that emitted them. Detection of such objects would provide exquisitely precise tests of Einstein's theory of gravity. There may also be waves from unanticipated or exotic sources, such as backgrounds produced during the earliest moments of the universe or cusps associated with cosmic strings.

7.5.2 International Pulsar Timing Array (IPTA)

Going to even lower frequencies, pulsar timing can be used to detect gravitational waves with wave periods of order years. As we learned in Chapter 4, a pulsar is a very good timekeeper. The best pulsar timekeepers are millisecond pulsars, not only are their pulses very short, they are very stable: unlike classical pulsars that occasionally glitch because of star quakes or because they suffer an accretion event. Pulsars have been monitored by radio telescopes since their discovery in 1968 (Hewish et al. 1968). The first indirect detection of gravitational waves was by observing a binary pulsar (Hulse & Taylor 1974). The best tests of GR in the strong-field were made using pulsar timing (Kramer et al. 2006). The detected pulses are shaped by a variety of physical effects: the pulsar's magnetosphere, the interstellar medium, and the properties of the detectors. In addition as pointed out by Hobbs and collaborators (Hobbs et al. 2010), the pulse arrival times will also be effected by any gravitational waves that pass through the pulsar or the detector.

The IPTA continuously monitors between 30 and 50 pulsars searching for anomalies, that, after taking into account all the factors that affect the time, search for gravitational waves as the cause. IPTA is a consortium of several individual PTAs: The European project (EPTA; Janssen et al. 2008; Ferdman et al. 2010), the North American Nanohertz Observatory for Gravitational Waves (NANOGrav; Demorest et al. 2013), and Parkes PTA (PPTA; Hobbs et al. 2009). The observed pulse arrival times are compared with models that predict the expected arrival times. The residuals between the model and observations contain the effects of the gravitational waves. While individual pulsars are very stable clocks, using a large number of them together allows for better accounting of the dispersion effects in the atmosphere and in the interstellar matter. Supermassive black holes are thought to be the main source of the background gravitational waves that IPTA hopes to detect.

References

Aab, A., Abreu, P., Aglietta, M., et al. 2016, PhRvL, 117, 192001
Aartsen, M. G., Ackermann, M., Adams, J., et al. 2019, ApJ, 886, 12
Abbott, B. P., Abbott, R., & Abbott, T. D. 2016, ApJ, 818, 22
Ambrosi, G., An, Q., Asfandiyarov, R., et al. 2017, Natur, 552, 63
An, Q., Asfandiyarov, R., Azzarello, P., et al. 2019, SciA, 5, eaax3793
Babcock, H. 1953, PASP, 65, 194
Caiazzo, I., & Heyl, J. 2021a, MNRAS, 501, 109
Caiazzo, I., & Heyl, J. 2021b, MNRAS, 501, 129
Demorest, P. B., Ferdman, R. D., Gonzalez, M. E., et al. 2013, ApJ, 762, 94
Ferdman, R. D., van Haasteren, R., Bassa, C. G., et al. 2010, CQGra, 27, 084014
Hewish, A., Bell, S. J., Pilkington, J. D. H., Scott, P. F., & Collins, R. A. 1968, Natur, 217, 709
Heyl, J. 2022, Univ, 8, 84
Hobbs, G. B., Bailes, M., Bhat, N. D. R., et al. 2009, PASA, 26, 103
Hobbs, G., Archibald, A., Arzoumanian, Z., et al. 2010, CQGra, 27, 084013
Hulse, R. A., & Taylor, J. H. 1974, ApJ, 191, L59
Janssen, G. H., Stappers, B. W., Kramer, M., et al. 2008, A&A, 490, 753
Kramer, M., Stairs, I. H., Manchester, R. N., et al. 2006, AnP, 15, 34
Liu, S.-F., Jin, S., Li, S., Isella, A., & Li, H. 2018, ApJ, 857, 87
Meszaros, P., Fox, D. B., Hanna, C., & Murase, K. 2019, NatRP, 1, 585
Vadawale, S. V., Chattopadhyay, T., Mithun, N. P. S., et al. 2018, NatAs, 2, 50
Weekes, T. C., Cawley, M. F., Fegan, D. J., et al. 1989, ApJ, 342, 379
Weisskopf, M. C., Cohen, G. G., Kestenbaum, H. L., et al. 1976, ApJ, 208, 125
Weisskopf, M. C., Ramsey, B., O'Dell, S. L., et al. 2016, ResPh, 6, 1179
Zhang, K., Bergin, E. A., Williams, J. P., Pinilla, P., & Andrews, S. M. 2018, in ASP Conf. Ser. 7, ed. E. Murphy (San Francisco, CA: ASP), 209

www.ingramcontent.com/pod-product-compliance
Ingram Content Group UK Ltd.
Pitfield, Milton Keynes, MK11 3LW, UK
UKHW061857100926
8556IPUK00008B/911

9 780750 337298